Manuel R. Guariguata and Pedro H.S. Brancalion (Eds.)

Governing Forest Restoration: Social, Environmental and Institutional Dimensions

MDPI

This book is a reprint of the special issue that appeared in the online open access journal *Forests* (ISSN 1999-4907) in 2014 (available at: http://www.mdpi.com/journal/forests/special_issues/governing_forest).

Guest Editors
Manuel R. Guariguata
Center for International Forestry Research
Perú

Pedro H.S. Brancalion
Department of Forestry, "Luiz de Queiroz" College of Agriculture
University of São Paulo, Piracicaba
SP, Brazil

Editorial Office
MDPI AG
Klybeckstrasse 64
Basel, Switzerland

Publisher
Shu-Kun Lin

Assistant Managing Editor
Echo Zhang

1. Edition 2015

MDPI • Basel • Beijing • Wuhan

ISBN 978-3-03842-042-2

Table of Contents

List of Contributors

Lucy Aquino: WWF Paraguay, Avenida Argaña c/ Peron, Edificio Opa Rudy, 4to Piso, Asuncion, Paraguay

Alaine A. Ball: Yale University MacMillan Center for International and Area Studies, Henry R. Luce Hall, 34 Hillhouse Avenue, P.O. Box 208206, New Haven, CT 06520-8206, USA; Departamento de Ciências Florestais, Escola Superior de Agricultura "Luiz de Queiroz", Universidade de São Paulo, Avenida Pádua Dias 11, 13418-260, Piracicaba-SP, Brazil

Jos Barlow: Museu Paraense Emilio Goeldi, Av. Magalhães Barata, 376, Belém, Pará, 66.040-170, Brazil, Lancaster Environment Centre, Lancaster University, Lancaster LA1 4YQ, UK

Michael T. Bennett: College of Environmental Sciences and Engineering, Peking University, Beijing 100871, China

Pedro H.S. Brancalion: Department of Forest Sciences, "Luiz de Queiroz" College of Agriculture, University of São Paulo, Avenida Pádua Dias 11, Piracicaba, SP 13418-260, Brazil

Miguel Calmon: Global Forest and Climate Change Programa (GFCCP), International Union for Conservation of Nature, Washington, DC, USA

Helena Carrascosa: Secretaria do Meio Ambiente, São Paulo, SP, Brazil

Pedro Castro: Secretaria Executiva do Pacto pela Restauração da Mata Atlântica, Rua do Horto, 931, Casa das Reservas da Biosfera, São Paulo, SP, Brazil

Ricardo Gomes César: Laboratory of Tropical Forestry (LASTROP), Department of Forest Sciences, ESALQ—University of São Paulo, Av. Pádua Dias,11, P.O. Box 9, 13418-900 Piracicaba, São Paulo, Brazil

Guillaume de Buren: Swiss Graduate School of Public Administration (IDHEAP), University of Lausanne, Quartier UNIL-Mouline, Lausanne 1015, Switzerland

Carlos Alberto de Mattos Scaramuzza: Secretaria de Biodiversidade e Florestas, Ministério do Meio Ambiente (MMA), Distrito Federal, Brasília, DF, Brazil

Thomas K. Erdmann: Environment, Climate Change & Urban Services, DAI Solutions, 7600 Wisconsin Avenue, Suite 200, Bethesda, MD 20814, USA

Joice Ferreira: Embrapa Amazônia Oriental, Travessa Dr. Enéas Pinheiro s/n, CP 48, Belém, Pará, 66.095-100, Brazil

Horst Freiberg: German Federal Ministry for the Environment, Nature Conservation and Nuclear Safety (BMUB), Robert-Schuman-Platz 3, Bonn 53175, Germany

Toby Gardner: Stockholm Environment Institute, Linnégatan 87 D, Box 24218, Stockholm, 10451, Sweden, Instituto International para Sustentabilidade, Estrada Dona Castorina, 124, Horto, Rio de Janeiro, RJ 22.460-320, Brazil

Alice Gouzerh: Departamento de Ciências Florestais, Escola Superior de Agricultura "Luiz de Queiroz", Universidade de São Paulo, Avenida Pádua Dias 11, 13418-260, Piracicaba-SP, Brazil; Montpellier SupAgro, 2 place Pierre Viala, 34060 Montpellier Cedex 2, France
Departamento de Ciências Florestais, Escola Superior de Agricultura "Luiz de Queiroz", Universidade de São Paulo, Avenida Pádua Dias 11, 13418-260, Piracicaba-SP, Brazil

Manuel R. Guariguata: Center for International Forestry Research (CIFOR), Av. La Molina, Lima 12, Peru

Petrus Gunarso: Tropenbos International, P.O. Box 494, Balikpapan 76100, East Kalimantan, Indonesia

Nicholas J. Hogarth: Center for International Forestry Research (CIFOR), Bogor 16115, Indonesia

Habtemariam Kassa: Center for International Forestry Research, Forests and Livelihoods Research, CIFOR Ethiopia Office, P.O. Box 5689, Addis Ababa, Ethiopia

Irene Koesoetjahjo: Tropenbos International, P.O. Box 494, Balikpapan 76100, East Kalimantan, Indonesia

Renaud Lapeyre: Institute for Sustainable Development and International Relations (IDDRI), 41 rue du Four, Paris 75006, France

Alexander C. Lees: Museu Paraense Emilio Goeldi, Av. Magalhães Barata, 376, Belém, Pará, 66.040-170, Brazil

Mulugeta Lemenih: Forestry Department, Farm Africa, Ethiopia office, Addis Ababa, Ethiopia

Samuel I. Levy-Tacher: El Colegio de la Frontera Sur, ECOSUR, San Cristóbal de las Casas, Chiapas 29290, Mexico

Pedro Macario-Mendoza: El Colegio de la Frontera Sur, ECOSUR, Chetumal, Quintana Roo 77900, Mexico

Stephanie Mansourian: Mansourian Environmental Consultancy, 36 Mont d'Eau du Milieu, 1276 Gingins, Switzerland

Felipe Melo : Departamento de Botânica, Universidade Federal de Pernambuco, Av. Prof. Moraes Rego s/n, Recife, PE, Brazil

Carlos A. Mesquita : Diretor para a Mata Atlântica da Conservação Internacional, doutorando do Laboratório de Gestão Ambiental, Programa de Pós-Graduação em Ciências Ambientais e Florestais, Universidade Federal Rural do Rio de Janeiro (UFRRJ), Brazil

Aurélio Padovesi : The Nature Conservancy, Avenida Paulista, São Paulo, SP, Brazil

Daoli Peng: Beijing Forestry University (BFU), Beijing 100083, China

Francisco Pereira: WWF Paraguay, Avenida Argaña c/ Peron, Edificio Opa Rudy, 4to Piso, Asuncion, Paraguay

Severino R. Pinto: Centro de Pesquisas Ambientais do Nordeste (Cepan), Rua Dom Pedro Henrique, 167, Recife, PE, Brazil; Departamento de Botânica, Universidade Federal de Pernambuco, Av. Prof. Moraes Rego s/n, Recife, PE, Brazil

Romain Pirard: Center for International Forestry Research, Jalan CIFOR, Situ Gede, Bogor Barat 16115, Indonesia

Till Pistorius: Forest and Environmental Policy Group, University of Freiburg, Tennenbacherstr. 4, Freiburg im Breisgau D-79106, Germany

Louis Putzel: Center for International Forestry Research (CIFOR), Bogor 16115, Indonesia

Ricardo Rodrigues: Laboratory of Forest Ecology and Restoration (LERF), Department of Biological Sciences, ESALQ, University of São Paulo, Av. Pádua Dias,11, P.O. Box 9, 13418-900 Piracicaba, São Paulo, Brazil

Francisco J. Román-Dañobeytia: Center for Latin American Studies, University of Florida, Gainesville, FL 32611, USA, Madre de Dios Consortium, Puerto Maldonado, Madre de Dios 17001, Peru

Marcelo Tabarelli : Departamento de Botânica, Universidade Federal de Pernambuco, Av. Prof. Moraes Rego s/n, Recife, PE, Brazil

Cora van Oosten: Wageningen UR Centre for Development Innovation, Wageningen University & Research, Wageningen, P.O. Box 88, 6700 AB Wageningen, The Netherlands

Ima Célia Guimarães Vieira: Museu Paraense Emilio Goeldi, Av. Magalhães Barata, 376, Belém, Pará, 66.040-170, Brazil

Freerk Wiersum: Forest and Nature Conservation Policy Group, Wageningen University, Wageningen, P.O. Box 47, 6700 AA Wageningen, The Netherlands

Chen Xie: China National Forestry Economics and Development Research Center (FEDRC), Beijing 100714, China

José Zúñiga-Morales: Comisión de Áreas Naturales Protegidas, Reserva de la Biósfera Calakmul, Zoh-Laguna, Calakmul, Campeche 24644, Mexico

Editorial

Current Challenges and Perspectives for Governing Forest Restoration

Manuel R. Guariguata and Pedro H. S. Brancalion

Abstract: Negotiation, reconciliation of multiple scales through both ecological and social dimensions and minimization of power imbalances are considered critical challenges to overcome for effective governance of forest restoration. Finding the right mix of "command and control" in forest restoration *vs.* "environmental governance", which includes non-state actors, regulatory flexibility, and market based instruments is at the heart of these challenges. This Special Issue attempts at shedding light on these challenges with case studies from South and Central America, Africa, and Asia. Some provide within-country as well as cross-country comparisons. A few others present case studies at the household level. Both policy and legal constraints towards implementing forest restoration are also discussed as a function of top down *vs.* bottom up approaches. The effectiveness of payments for environmental services is examined as catalyzers of forest restoration initiatives. Finally, two papers deal with the legal and policy constraints in making restoration through natural regeneration a viable and cost-effective tool. In the face of renewed perspectives for expanding forest restoration programs globally, governance issues will likely play a key role in eventually determining success. As many of the papers in this Special Issue suggest, the fate of forest restoration outcomes is, more often than not, associated with overall governance challenges, some of which are often overlooked particularly across multiple scales.

Reprinted from *Forests*. Cite as: Guariguata, M.R.; Brancalion, P.H.S. Current Challenges and Perspectives for Governing Forest Restoration. *Forests* **2014**, *5*, 3022-3030.

1. Introduction

The main forest governance trends that defined the last two decades are thought to be: (i) decentralization of management; (ii) the role of industrial logging in government-granted forest concessions; and (iii) market-oriented certification [1]. Furthermore, it is also agreed that the following key variables influence the outcomes of forest governance particularly across the tropics: (i) user rights and responsibilities; (ii) participation by those who use and depend on forests; (iii) accountability of decision-makers; (iv) monitoring of forest management; (v) enforcement of property rights; and (vi) institutional capacities [1]. Yet for the most part, these conclusions have emerged when the natural resource base is already sufficient for providing forest-based goods and services in the long-term [2], in other words, management of natural forests. Less seems to be known when the objective is restoring forest cover in degraded or otherwise deforested land. While there is no major reason to believe that the abovementioned issues would not also shape forest restoration outcomes (see examples in [3]), the challenges may differ from those applied to natural forests.

First, forest restoration not only deals with access and/or use of a novel resource base but also with long lasting changes in land use allocation and human use. As a consequence, if forest restoration is not planned, implemented, and managed to eventually become a profitable and equitable land use option [4], conflicts may arise [5]. Second, it is only recently that the social sciences have been included in theoretical frameworks and the practice of ecological restoration [6,7] along with the insertion of historical, political, judicial, aesthetic, as well as moral issues [8–10]. That said, only a decade ago Dudley *et al.* [11] urged practitioners to move beyond tree planting, to restore with a landscape mindset while considering both biophysical and socioeconomic issues, and to insert the views of different stakeholders and institutions. Governing "forest landscape restoration" could be seen as a nascent field with methodological, conceptual, and practical challenges ahead.

Yet with the growing recognition that forest restoration offers great opportunities for supporting biodiversity conservation and ecosystem services provisioning at the global and local levels [12], the number of projects have dramatically increased along with their spatial scale [13–16]. Particularly in a landscape context, negotiation, reconciliation of multiple scales through both ecological and social dimensions and minimization of power imbalances are considered critical challenges to overcome [17]. Finding the right mix of "command and control" in forest restoration *vs.* "environmental governance" [18], which includes non-state actors, regulatory flexibility, and market based instruments, is at the heart of these challenges. Governance systems may therefore need to be adapted to include a wide range of stakeholders, legal instruments, inter-sectorial policies, and multi-level government administrations. Here we apply the definition of governance used by Colfer and Pfund [19]: "The ways and institutions through which individuals and groups express their interests, exercise the rights and obligations, and mediate their differences."

2. The Contents of the Special Issue

The papers composing this Special Issue attempt to shed some light on the abovementioned issues. In total, the 10 papers cover case studies in seven countries, from South and Central America, Africa, and Asia. They cover both the tropics and subtropics and include global, national and local scales. Some provide within-country as well as cross-country comparisons, while others focus on natural regeneration as a mode of forest restoration. A few others present case studies at the household level. Both policy and legal constraints towards implementing forest restoration are also discussed as a function of "top down" *vs.* "bottom up" approaches. The effectiveness of payments for environmental services is examined as catalyzers of forest restoration initiatives.

Forest restoration has been a prominent topic in the agenda of many international forums over the last decade from a climate, biodiversity, desertification and sustainable development perspective (reviewed in Lamb [3]). In particular, at the 2010 Conference of the Parties to the Convention on Biological Diversity, two ambitious proposals (Aichi Targets 14 and 15) were adopted with the aim of restoring degraded land at a global scale. The article by Pistorius and Freiberg [20] focuses on the international restoration policy arena and discusses the major challenges facing the mobilization of billions of US dollars that may be needed to reach these targets. It is estimated that 20 million ha of terrestrial degraded land will have to be restored per year until 2020. The authors discuss how a "collaborative governance approach" may be needed for an effective implementation of these targets, since they conclude that the current global institutional landscape is too fragmented. More often than not, international targets, although representing genuine aspirations, do not permeate down to national and local levels. That said, Pistorius and Freiberg make the case for serious consideration of multi-stakeholder partnerships (between the private-and public sectors, and civil society) as one way forward.

The issue of collaborative governance to overcome institutional fragmentation is further discussed at the local level in the paper by Pinto *et al.* [21]. Although in many countries, particularly across the tropics, government-led reforestation and restoration programs are the norm, the authors make the case against top down approaches and the lack of positive incentives in the practice of forest restoration by presenting the governance structure of the Atlantic Forest Restoration Pact (AFRP) in Brazil. Fragmented efforts and disregard of critical bottlenecks for upscaling plot-based restoration prompted the development of a multistakeholder and interdisciplinary platform where different interests, perspectives, skills and approaches converge towards a common goal along with the ambitious target to restore 15 m ha of the Atlantic Forest biome. One important lesson learned from the AFRP mechanism is that solely relying on legal compliance for implementing restoration is neither sufficient nor desirable. Note that in many cases, restoration governance is dictated by top-down legal instruments, such as biodiversity offsets [22]. In fact, nearly 60% of the studies reviewed by Ruiz-Jaen and Aide [23] were carried out to comply with environmental laws.

The interactions (or lack thereof) among different actors other than government; of voluntary and negotiated agreements; of flexible approaches to negotiated implementation; and of market-based instruments are documented across the papers described ahead. In Ethiopia, Lemenih and Kassa [24] argue that although ambitious reforestation and restoration targets were set by the central government (2 million ha for afforestation and 1 million ha for reforestation), it has been mainly non-governmental organizations (NGOs) that have played a key role in their implementation while advocating for policy reforms. Yet the authors point out that lack of knowledge sharing among NGOs has resulted at times in contradictory messages to both local communities and policy makers. The authors further conclude that at present, the various reforestation and restoration practices in Ethiopia lack coordination, both technically and managerially and also lack the application of indicators of performance for measuring success. Also at the country level, but through a comparative view, the paper by van Oosten *et al.* [25] from Indonesia analyzes how people's views and participation are inserted in those landscapes to be restored in three contrasting situations: (i) the extension of a national park; (ii) compliance with environmental law; and (iii) collective action. Three case studies represent different interpretations on the nature of foreste restoration, different governance mechanisms and different extent of stakeholder involvement. In each case, flexible governance arrangements were lacking from the outset and therefore institutional space for negotiated decision-making had to be created. In Brazil, by using a case study from São Paulo State, Ball *et al.* [26] describe lessons learned on how a landscape-scale forest restoration project was conceived by an international NGO in the Atlantic Forest biome. The project targeted small-scale landholders to restore their degraded lands by offering restoration models that would provide economic benefits. In spite of initial optimism, problems in the implementation phase arose. These related to fund allocation, legal regulations hampering native species harvesting, and on the adequate integration of the needs and perspectives of participants. The authors recommend baseline social assessments to improve project design, simplification of legal frameworks to exploit native species, and better communication and articulation among stakeholders.

In the only paper of the Special Issue having a cross-country approach, Mansourian *et al.* [27] explore how different governance challenges are displayed under different forest tenure arrangements in private forests in Paraguay and co-managed forests in Madagascar. Two key factors raised as necessary for effective and equitable forest landscape restoration are: (i) improving the forest governance context so that processes are more effective and key stakeholder groups can increase their participation in restoration activities; and (ii) promoting positive incentives for implementing restoration including compensation for the provision of ecosystem services. The authors also argue that in these two countries, fragmented multilevel governance and poor policy-making further hinder forest landscape restoration. In Paraguay, what is seen as complex forest legislation does not seem to parallel the level of support needed by institutions that are to implement and enforce such legislation. While in Madagascar, the main reason for local-level engagement in co-management arrangements is likewise a response to what are seen as defective policies regarding management and ownership rights.

Previous analyses have underscored the potential for forest restoration to enhance the delivery of environmental services for global and local benefits [28]. In the context of positive incentives through conditional payments, two papers shed relevant information on this topic. Pirard *et al.* [29] studied two watershed restoration projects in Indonesia, both of which are assumed to increase dry season watershed flows (through tree planting), and the concurrent ability of payment for environmental services (PES) schemes to improve the effectiveness of these initiatives (compared to government-led watershed programs). The authors conclude that despite their innovations over command and control approaches, the applied PES schemes have had limited effectiveness in promoting forest restoration. However the ability of local stakeholders in adapting to changes in the way these programs have evolved over time (after the intervention of several international actors) has generated a sense of collective ownership towards the goal of securing water provision. The paper by Bennett *et al.* [30] refer to what is known as the world's largest afforestation-based payments for ecosystem services program (27 million ha), China's Conversion of Cropland to Forest Program (CCFP). The authors examined the factors associated with the survival rate of planted seedlings, which is used by CCFP both as a measure of the impact of program incentives and to deliver subsidies to participating rural households. A key result is that households with higher levels of human capital in forestry activities appear to be better at keeping trees alive. Another key result is that the degree to which local governments engage with participants during program implementation has a positive effect on program outcomes.

When land use after deforestation and degradation has been neither heavy nor prolonged, forest ecosystems are known to recover rapidly without human intervention through second-ary succession [31]. Thus relying on natural regeneration is a viable restoration approach when carefully assessed against those variables known to define both the speed and trajectory of unassisted forest recovery [32]. Yet for decades, secondary forests have remained "under the radar" in both national and international land use planning strategies [33] in spite of inter-national efforts to recognize these forests as a legitimate land use type [34]. Interestingly, the legal frameworks governing second-growth forests in many tropical countries are frequently marked by ambiguity (e.g., Sears *et al.* [35]) while the opposite applies for human-assisted forest restoration where public policy and detailed legislation is very clear [10]. Two papers in this Special issue deal with the legal and policy constraints in making natural regeneration a cost-effective tool while satisfying both the needs of local people and conservation objectives.

The paper by Vieira *et al.* [36] analyzes the key legal impediments facing the development of a system of good governance for second-growth forests in the state of Pará, in the Brazilian Eastern Amazon. In contrast to the rest of Brazil where in most states there is no legal defini-tion for when a regenerating area becomes classified as "forest" rather than "fallow" (and thus qualifies for legal protection), Pará is the only state of the Brazilian Amazon that has adopted an explicit definition of second-growth forests based on biophysical parameters (once the def-inition applies, these cannot be cleared in order to comply with conservation objectives). However, the authors discuss how effective governance of this widespread tropical resource is challenged by lack of clarity in terms and definitions, inconsistencies in legal frameworks from the federal to the local level, and an overall perception by society and policy makers that secondary forest ecosystems have little value. The authors further conclude that for secondary forests to restore ecological and social values through natural regeneration, management deci-sions should not be made based on technical indicators of forest condition alone but should incorporate an understanding of the drivers of success, encompassing the suite of inter-related biophysical, socio-economic and institutional factors. To this end, dialogue between the envi-ronmental and agricultural decision makers is warranted.

A related situation occurs in Mexico where secondary forests are defined strictly on bio-physical grounds. The paper by Román-Dañobeytia *et al.* [37] evaluated the relevance of the rigidly applied reference values in the current forestry law that defines what a secondary for-est is. They suggest that these standard values limit potential management actions in the Yu-catán Peninsula. In contrast to the case study from Brazil mentioned above, once the biophys-ical reference values are reached, the forest is subject to overregulation. In other words, sec-ondary forests are prevented by law from traditional, extractive uses, as formal management plans are required. As in the above case, cross-sectorial dialogue is needed.

3. Conclusions

Not long ago, and right after the first formal definitions of "ecological restoration" provid-ed by Bradshaw and Chadwick [38], the discipline developed somewhat in isolation. As seen in this Special Issue, restoring degraded or non-forested lands is an inherently multidisciplinary, multiscalar and multisectorial activity in need of good governance so that the rights and obligations as well as mediation of differences among stakeholders contribute both to achieving predetermined restoration objectives and the maintenance of the resource base (Figure 1). In the face of renewed perspectives for expanding forest restoration programs globally [39], governance issues will likely play a key role in eventually determining success. As many of the papers in this Special Issue suggest, the fate of forest restoration is frequently associated with systemic governance challenges, which are all too often overlooked. It is therefore hoped that this Special Issue provides at least some useful background for designing and implementing new and more effective forest restoration programs globally.

Figure 1. Key governance issues driving ecological and socioeconomic processes associated to land and forest states modification and their outcomes.

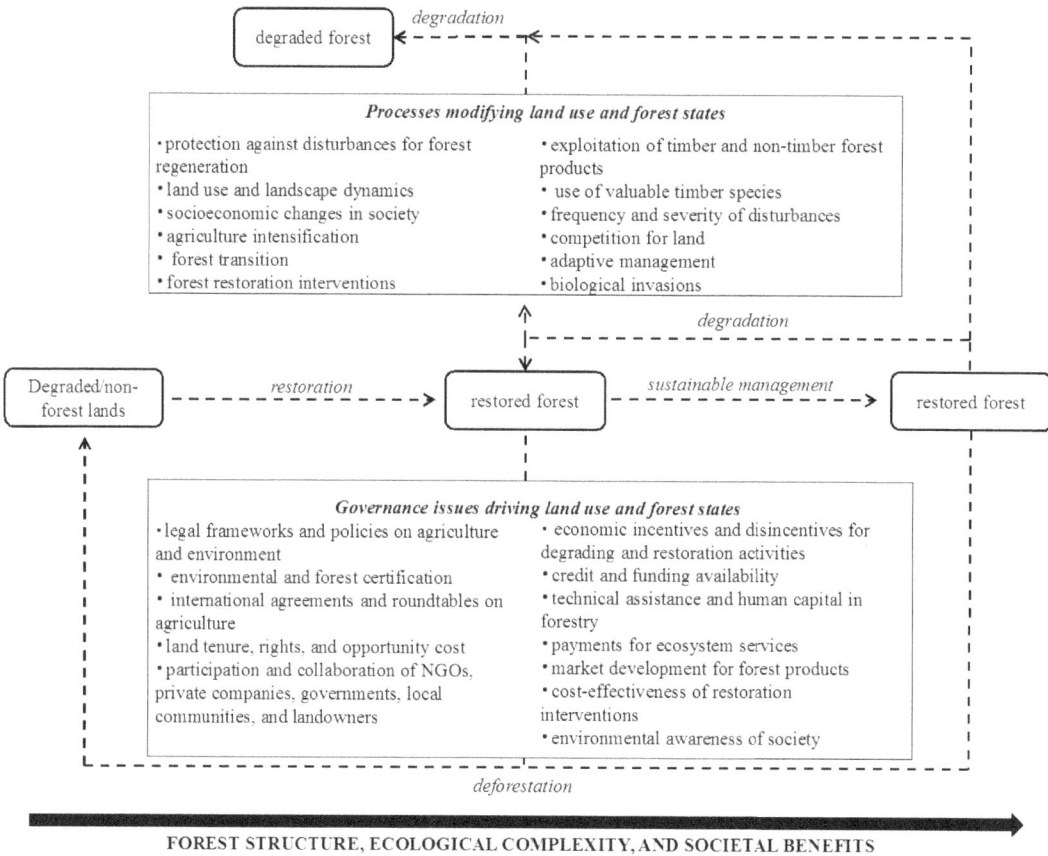

Processes modifying land use and forest states

- protection against disturbances for forest regeneration
- land use and landscape dynamics
- socioeconomic changes in society
- agriculture intensification
- forest transition
- forest restoration interventions

- exploitation of timber and non-timber forest products
- use of valuable timber species
- frequency and severity of disturbances
- competition for land
- adaptive management
- biological invasions

Governance issues driving land use and forest states

- legal frameworks and policies on agriculture and environment
- environmental and forest certification
- international agreements and roundtables on agriculture
- land tenure, rights, and opportunity cost
- participation and collaboration of NGOs, private companies, governments, local communities, and landowners

- economic incentives and disincentives for degrading and restoration activities
- credit and funding availability
- technical assistance and human capital in forestry
- payments for ecosystem services
- market development for forest products
- cost-effectiveness of restoration interventions
- environmental awareness of society

FOREST STRUCTURE, ECOLOGICAL COMPLEXITY, AND SOCIETAL BENEFITS

Acknowledgments

We thank Stephanie Mansourian for editing an earlier draft.

Author Contributions

Both authors wrote together the article presented here.

Conflicts of Interest

The authors declare no conflict of interest.

Manuel R. Guariguata and Pedro H.S. Brancalion
Guest Editors

References

1. Agrawal, A.; Chhatre, A.; Hardin, R. Changing governance of the world's forests. *Science* **2008**, *320*, 1460–1462.
2. *Forests under Pressure: Local Responses to Global Issues*; IUFRO World Series; Katila, P., Galloway, G., de Jong, W., Pacheco, P., Mery, G., Eds.; IUFRO: Vienna, Austria, 2014; Volume 32.
3. Lamb, D. *Large-Scale Forest Restoration*; Earthscan-Routledge: London, UK, 2014.
4. Brancalion, P.H.S.; Viani, R.A.G.; Strassburg, B.B.N.; Rodrigues, R.R. Finding the money for tropical forest restoration. *Unasylva* **2012**, *63*, 41–50.
5. Knoke, T.; Calvas, B.; Aguirre, N.; Román-Cuesta, R.M.; Günter, S.; Stimm, B.; Weber, M.; Mosandl, R. Can tropical farmers reconcile subsistence needs with forest conservation? *Front. Ecol. Environ.* **2009**, *7*, 548–554.
6. Jackson, L.L.N.; Lopukine, N.; Hillyard, D. Ecological restoration: A definition and comments. *Restor. Ecol.* **1995**, *3*, 71–75.
7. Higgs, E.S. What is good ecological restoration? *Conserv. Biol.* **1997**, *11*, 338–348.
8. Clewell, A.F.; Aronson, J. *Ecological Restoration: Principles, Values, and Structure of an Emerging Profession*; Island Press: Washington, DC, USA, 2013. Second Edition.
9. Armesto, J.J.; Bautista, S.; del Val, E.; Ferguson, B.; García, X.; Gaxiola, A.; Godinez-Álvarez, H.; Gann, G.; López-Barrera, F.; Manson, R.; *et al.* Towards an ecological restoration network: Reversing land degradation in Latin America. *Front. Ecol. Environ.* **2007**, *5*, w1–w4.

10. Aronson, J.; Brancalion, P.H.S.; Durigan, G.; Rodrigues, R.R.; Engel, V.L.; Tabarelli, M.; Torezan, J.M.D.; Gandolfi, S.; Melo, A.C.G.; Kageyama, P.Y.; *et al*. What role should government regulation play in ecological restoration? Ongoing debate in São Paulo State, Brazil. *Restor. Ecol.* **2011**, *19*, 690–695.

11. Dudley, N.; Mansourian, S.; Vallauri, D. Forest landscape restoration in context. In *Forest Restoration in Landscapes: Beyond Planting Trees*; Mansourian, S., Vallauri, D., Dudley, N., Eds.; Springer: New York, NY, USA, 2005; pp. 3–7.

12. Aronson, J.; Alexander, S. Ecosystem restoration is now a global priority: Time to roll up our sleeves. *Restor. Ecol.* **2013**, *21*, 293–296.

13. McQueen, C.; Noemdoe, S.; Jezile, N. The Working for Water Programme. *Land Use Water Resour. Res.* **2001**, *1*, 1–4.

14. Doyle, M.; Drew, C.A. *Large-Scale Ecosystem Restoration Five Case Studies from the United States*; Island Press: Washington, DC, USA, 2008.

15. Arriagada, R.A.; Ferraro, P.J.; Sills, E.O.; Pattanayak, S.K.; Cordero-Sancho, S. Do payments for environmental services affect forest cover? A farm-leve evaluation from Costa Rica. *Land Econ.* **2012**, *88*, 382–399.

16. Calmon, M.; Brancalion, P.H.S.; Paese, A.; Aronson, J.; Castro, P.; Silva, S.C.; Rodrigues, R.R. Emerging threats and opportunities for large-scale ecological restoration in the Atlantic Forest of Brazil. *Restor. Ecol.* **2011**, *19*, 154–158.

17. Kozar, R.; Buck, L.E.; Barrow, E.G.; Sunderland, T.C.H.; Catacutan, D.E.; Planicka, C.; Hart, A.K.; Willemen, L. *Toward Viable Landscape Governance Systems: What Works?* EcoAgriculture Partners: Washington, DC, USA, 2014.

18. Lemos, M.C.; Agrawal, A. Environmental governance. *Annu. Rev. Environ. Resour.* **2006**, *31*, 297–325.

19. Colfer, C.J.P.; Pfund, J.L. *Collaborative Governance of Tropical Landscapes*; Earthscan: London, UK, 2011.

20. Pistorius, T.; Freiberg, H. From target to implementation: Perspectives for the international governance of forest landscape restoration. *Forests* **2014**, *5*, 482–497.

21. Pinto, S.R.; Melo, F.; Tabarelli, M.; Padovesi, A.; Mesquita, C.A.; Scaramuzza, C.A.M.; Castro, P.; Carrascosa, H.; Calmon, M.; Rodrigues, R.R.; *et al*. Governing and delivering a biome-wide restoration initiative: The case of Atlantic Forest Restoration Pact in Brazil. *Forests* **2014**, *5*, 2212–2229.

22. Maron, M.; Hobbs, R.J.; Moilanen, A.; Matthews, J.W.; Christie, K.; Gardner, T.A.; Keith, D.A.; Lindenmayer, D.B.; McAlpine, C.A. Faustian bargains? Restoration realities in the context of biodiversity offset policies. *Biol. Conserv.* **2012**, *155*, 141–148.

23. Ruiz-Jaen, M.C.; Aide, T.M. Restoration success: How is it been measured? *Restor. Ecol.* **2005**, *13*, 569–577.

24. Lemenih, M.; Kassa, H. Re-greening Ethiopia: History, challenges and lessons. *Forests* **2014**, *5*, 1896–1909.

25. Van Oosten, C.; Gunarso, P.; Koesoetjahjo, I.; Wiersum, F. Governing forest landscape restoration: Cases from Indonesia. *Forests* **2014**, *5*, 1143–1162.

26. Ball, A.; Gouzerh, A.; Brancalion, P.H.S. Multi-Scalar governance for restoring the Brazilian Atlantic Forest: A case study on small landholdings in protected areas of sustainable development. *Forests* **2014**, *5*, 599–619.

27. Mansourian, S.; Aquino, L.; Erdmann, T.K.; Pereira, F. A Comparison of governance challenges in forest restoration in Paraguay's privately-owned forests and Madagascar's co-managed state forests. *Forests* **2014**, *5*, 763–783.

28. Alexander, S.; Nelson, C.R.; Aronson, J.; Lamb, D.; Cliquet, A.; Erwin, K.L.; Finlayson, C.M.; de Groot, R.S.; Harris, J.A.; Higgs, E.S.; *et al.* Opportunities and challenges for ecological restoration within REDD+. *Restor. Ecol.* **2011**, *19*, 683–689.

29. Pirard, R.; Buren, G.; Lapeyre, R. Do PES improve the governance of forest restoration? *Forests* **2014**, *5*, 404–424.

30. Bennett, M.T.; Xie, C.; Hogarth, N.J.; Peng, D.; Putzel, L. China's Conversion of Cropland to Forest Program for household delivery of ecosystem services: How important is a local implementation regime to survival rate outcomes? *Forests* **2014**, *5*, 2345–2376.

31. Chazdon, R.L. Tropical forest recovery: Legacies of human impact and natural disturbances. *Perspect. Plant Ecol.* **2003**, *6*, 51–71.

32. Holl, K.D.; Aide, T.M. When and where to actively restore ecosystems? *For. Ecol. Manag.* **2011**, *261*, 1558–1563.

33. Davies, P. La visibilidad de los bosques secundarios. In *Memorias del Taller Internacional Sobre el Estado Actual y Potencial de Manejo y Desarrollo de Bosque Secundario Tropical en América Latina*; Tratado de Cooperación Amazónica: Caracas, Venezuela, 1997.

34. ITTO. *ITTO Guidelines for the Restoration, Management and Rehabilitation of Degraded and Secondary Tropical Forests*; ITTO Policy Development Series, No. 13; ITTO: Yokohama, Japan, 2011.

35. Sears, R.; Cronkleton, P.; Perez-Ojeda, M.; Robiglio, V.; Putzel, L.; Cornelius, J. *Producción de Madera en Sistemas Agroforestales de Pequeños Productores*; Center for International Forestry Research: Bogor, Indonesia, 2014.

36. Vieira, I.C.G.; Gardner, T.G.; Ferreira, J.; Lees, A.C.; Barlow, J. Challenges of governing second-growth forests: A case study from the Brazilian Amazonian State of Pará. *Forests* **2014**, *5*, 1737–1752.

37. Román-Dañobeytia, F.J.; Levy-Tacher, S.I.; Macario-Mendoza, P.; Zúñiga-Morales, J. Redefining secondary forests in the Mexican Forest Code: Implications for management, restoration, and conservation. *Forests* **2014**, *5*, 978–991.

38. Bradshaw, A.D.; Chadwick, M.J. *The Restoration of Land: The Ecology and Reclamation of Derelict and Degraded Land*; Blackwell Publishing: London, UK, 1980.

39. Menz, M.H.M.; Dixon, K.W.; Hobbs, R.J. Hurdles and opportunities for landscape-scale restoration. *Science* **2013**, *339*, 526–527.

Do PES Improve the Governance of Forest Restoration?

Romain Pirard, Guillaume de Buren and Renaud Lapeyre

Abstract: Payments for Environmental Services (PES) are praised as innovative policy instruments and they influence the governance of forest restoration efforts in two major ways. The first is the establishment of multi-stakeholder agencies as intermediary bodies between funders and planters to manage the funds and to distribute incentives to planters. The second implication is that specific contracts assign objectives to land users in the form of conditions for payments that are believed to increase the chances for sustained impacts on the ground. These implications are important in the assessment of the potential of PES to operate as new and effective funding schemes for forest restoration. They are analyzed by looking at two prominent payments for watershed service programs in Indonesia—Cidanau (Banten province in Java) and West Lombok (Eastern Indonesia)—with combined economic and political science approaches. We derive lessons for the governance of funding efforts (e.g., multi-stakeholder agencies are not a guarantee of success; mixed results are obtained from a reliance on mandatory funding with ad hoc regulations, as opposed to voluntary contributions by the service beneficiary) and for the governance of financial expenditure (e.g., absolute need for evaluation procedures for the internal governance of farmer groups). Furthermore, we observe that these governance features provide no guarantee that restoration plots with the highest relevance for ecosystem services are targeted by the PES.

Reprinted from *Forests*. Cite as: Pirard, R.; de Buren, G.; Lapeyre, R. Do PES Improve the Governance of Forest Restoration? *Forests* **2014**, *5*, 404-424.

1. Introduction

This article studies the implications of innovative funding instruments for forest restoration, acknowledging that an increasing proportion of lands are degraded in the tropics [1] and that private and market-oriented approaches are acknowledged for their potential ability to address environmental issues [2]. Admittedly, a number of economic approaches for natural resource management as a whole have been in existence for a long time (e.g., with incentives provided through fiscal policies, see [3] for an ambitious analysis that embraces a multitude of instruments), and land degradation is not a new phenomenon. However, the scale has changed dramatically and effective policies are today required more than ever. The search for these policies for forest and land management is influenced by a pervasive context of discourses presenting environmental services [4] as the way forward [5]. Thus, Payments for Environmental Services (PES), a direct application of the latter concept combined with market-oriented approaches, became the subject of many experiments [6] and the center of attention of scientists, practitioners and policy-makers. It essentially involves voluntary payments by the beneficiaries of a service to its providers, so long as pre-agreed conditions are met, hence relying on individual incentives to account for externalities in land-use decisions.

There are many ways to study their implications for forest restoration [7], e.g., effectiveness [8,9], equity [10,11], sustainability of funding [12] or even the risks of disappearing intrinsic motivations for the preservation of nature [13,14], to single out only a few examples among a rapidly growing literature. In this article, we are interested in the implications specifically related to governance [15], which is the focus of this special issue. Governance refers here to the number, nature and interactions of the stakeholders that are involved in the programs, and to the institutional arrangements that are put in place for funding and spending among land users. It is therefore as much a matter of participation and local politics as it is a matter of technical arrangements to make sure that funding is sustained and spending leads to effective outcomes for land management.

Previous research has emphasized the risks and challenges of forest restoration in Asia-Pacific when based on large-scale governmental programs [16]. Taking a political economy approach, the authors identified a number of governance challenges that might impede an effective implementation of forest restoration initiatives. Among these they cite the control of state agencies and the political connections of the main corporate actors, the existence of corruption practices and ultimately the risk that reforestation activities prioritize lands with natural forest cover (and hence forest conversion before reforestation). They conclude that *"tree-planting programs have been guided by forest rent distribution practices of state forest bureaucracies and by corporate accumulation strategies"* (p. 9).

We add to this analysis by looking at reforestation and forest restoration efforts from a different angle, with a focus on small-scale and privately-funded experiments based on the PES rationale. The latter payment schemes are indeed presented by some as particularly effective when applied to restoration purposes [17,18]. On the one hand, PES schemes are reported to enable investors and practitioners to face high up-front capital needs and labor costs associated with tree plantations [19]. On the other hand, these schemes are assumed to provide farmers with technical assistance and economic incentives, which guarantee local participation in reforestation activities over time [20] and orient farmers' behavior towards forest restoration [19]. Besides, it is contended that PES will also be a critical new source of funding generated by public and private demand for ecosystem services [21] so as to financially support restoration activities [22,23].

Our study is a contribution to this debate about the compared merits of "traditional" *vs.* "innovative" approaches to forest restoration, from a governance perspective. It starts from the assumption that innovative instruments might provide better solutions for addressing the risks of embezzlement or corruption, as opposed to public programs, especially when the latter involve rent-seeking industrial corporate actors, as suggested by [16]. The distinction between these broad categories is somehow artificial and may not always be reflected by practice, but it still provides us with a starting point to conduct an investigation into the impacts that we can reasonably expect from any attempts to innovate funding and incentives in this domain.

The primary intent of our analysis is to answer the research question: "Do PES improve the governance of forest restoration programs as a basis for sustainable outcomes on the ground?" To investigate the impacts and added value of PES programs, we study the characteristics of their governance. This research question is addressed through the analysis of two assumptions: first, that a defining feature of PES compared to public programs is the key role given to multi-stakeholder agencies in terms of fund management, which is important from a governance perspective and creates the conditions for all views to be expressed including those of environmental NGOs and local residents; and second, that another crucial PES feature is the specific contracts that involve land users and assign objectives to them in the form of conditions for payments. These specific contracts result from service beneficiaries being attentive to effective service provision due to their direct, if not vital, interest in success, and they might provide more guarantees for sustained impacts on the ground.

A major problem for this analysis is the confusion around the term *PES* itself, and the diversity of understandings and experiments that the term encompasses. This "category" of policy instruments includes various types with contrasting characteristics, some of which are reported to match the characteristics of public subsidy programs [24]. This finding was further documented in [25], who made the point that many PES schemes could also be studied from the perspective of traditional public policies except for their underlying justification based on the remuneration of environmental services. Besides which, in many cases PES schemes tend to refer to the way that funding is secured for a given forest restoration initiative, notably through trust funds, rather than to the way that land users are involved through contracts [26]. Hence, we see that no black and white situation exists and the multidimensionality of all these policies and policy instruments tends to disqualify any attempts to make rigorous distinctions.

We have attempted to bypass these methodological hurdles in two ways. First, by studying two cases that illustrate the other end of the spectrum from public and national restoration programs, in that they are local and privately funded. Second, by looking at both sides of the table, namely funding (how financial resources are collected) and incentive distribution (how financial resources are spent). These two sides are complementary and involve governance challenges of equal importance for success. Funding determines the sustainability and scale of forest restoration efforts and can follow various paths from mandatory taxes to voluntary contributions; while incentive distribution determines the effectiveness of a scheme and can also take different forms, ranging from individually tailored contracts to flat subsidies.

Another source of confusion is the role of public authorities in "PES schemes". Clearly all depends on the scope of these policy instruments and what schemes this category encompasses. The evidence so far suggests that public authorities keep a firm grip and maintain a central role in many of the market-based instruments for environmental services, which runs counter to the common belief of a "rolling-back" of the state [27]. This fact also provides justification for our investigation and empirical documentation of the changes—if any—of governance induced by new mechanisms for forest restoration, which can certainly not be taken for granted.

Before proceeding with the analysis, we need to make one additional remark concerning the forest restoration activities that are studied in this article. While being justified by their positive contribution to water services, their actual effect on ground water is complex and controversial. The *"more trees more water"* myth is discussed and challenged in the literature [28–32], yet some recognize that forest cover might have positive impacts on infiltration in smaller scale catchments [33] with steep [34] and degraded soils [31,33]. Examples of improved groundwater storage are indeed documented in tropical forests [31,35]. The scope of our article is limited to the governance of a few restoration activities that are assumed by stakeholders to provide water services; it does not include a discussion of the impacts of forest cover on water. We only observe that forest restoration activities are undertaken based on their assumed capacity to increase the availability of groundwater in the dry season, which is an assumption that runs counter to some evidence in the literature [29,31,32].

In order to assess the reality of institutional changes in PES-related restoration schemes, we undertook field research in two of Indonesia's most prominent PES experiments, one in the Banten province west of Java (Cidanau) and the other in the island of Lombok in the eastern part of the country. The next section provides details about the chosen case studies and our analysis methods before presenting and discussing results.

2. Case Studies and Methods

2.1. Case Study Presentation

2.1.1. Lombok: Three Successive Funding Arrangements with Water Users

The first initiative is located on the island of Lombok in the eastern part of the archipelago and is one of the driest Indonesian islands. Its population is mainly concentrated in the lower plain where the capital city Mataram is located (see Figure 1), which has around 400,000 inhabitants. In the dry season, from March to October, there is little rainfall on the plain. The regional public water supply company (PDAM) therefore uses water catchments located at the bottom of the Rinjani volcano. These catchments play a key role in the regulation of water flows. However, a dramatic decrease in water flow from the springs was observed following the deforestation of the volcano's slopes in the 1990s, with around 50% of the springs drying up in the Rinjani area between 1985 and 2006, according to the Provincial environment agency (BLHP) [36]. Most stakeholders, both local and international, have assumed that these facts are related.

The whole process to develop a PES was initiated in 2001 with the financial and technical support of international agencies (US Agency for International Development (USAID), United Nations Development Programme (UNDP) and the Ford Foundation) through the organization of workshops and economic valuations addressing environmental issues in the watershed. These early activities led to a first and short-lasting PES experiment set up by local NGOs (Konsepsi, WWF-NT) and PDAM in the mid-2000s. Following a willingness-to-pay study among Mataram residents, an intermediary body (Bestari Community Funds) was created to collect and manage voluntary financial contributions. However, transaction costs were too high relative to the amount of money collected and potential to make a difference on the ground. Indeed there was no certainty about the available budget based on voluntary (hence, unpredictable and subject to large fluctuations and decline) contributions that remained extremely limited but with constant fixed costs to organize the system.

Figure 1. Location of the two case studies (Indonesia).

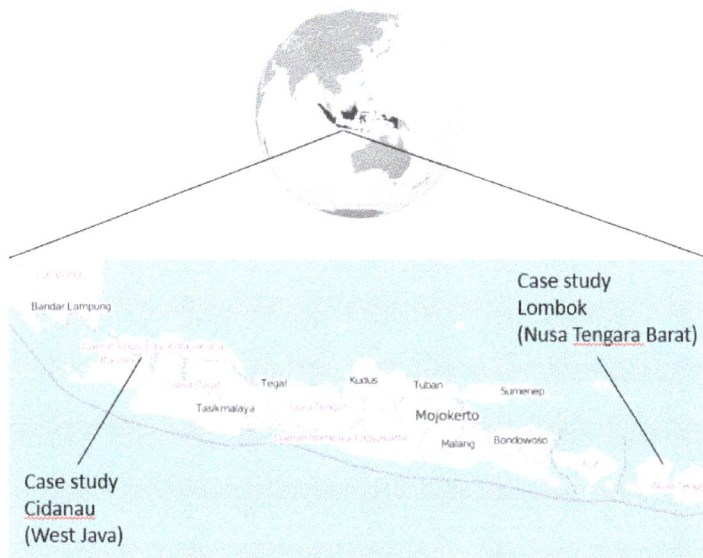

Since water tariffs are regulated by a regional decree, the district government had the opportunity to take control of the PES [37]. After a long legislative process, a new district regulation on "environmental services management" was issued in 2007, which paved the way for the establishment of a second PES from 2009 onwards. This second PES replaced the existing private intermediary body (Bestari Community Funds) with a sophisticated multi-stakeholder public agency (IMP). This new intermediary acted as fund manager, while implementing and controlling field operations, with the participation of civil society (WWF-NT, Konsepsi, *etc.*) and public agencies such as district authorities. The regulation established a monthly tax on water subscription that has been enforced since December 2009 and is collected through the PDAM billing system, and the funds have been used since 2010 by the IMP to cover expenses for forest restoration and local empowerment activities proposed by farmer groups. Restoration activities consist of the distribution of seedlings to individual farmers, under the supervision of a farmer group (kelompok).

Finally, a third PES scheme emerged in parallel after 2011, when the company PDAM (a major service beneficiary) decided to design and promote its own approach, probably because of the perceived ineffectiveness of the two previous attempts. This scheme involves bilateral agreements with farmers (without the multi-stakeholder agency IMP as intermediary) and takes place in parallel with the activities supported by the second scheme [37] (p. 272). Case selection is ultimately made by PDAM on the basis of proposals from the district forest service that in turn considers initial requests that originate in the farmer groups. While the funded activities (seedling distribution) that serve as incentives are very similar to the second and third schemes, the funding and coordination aspects are contrasting. Fee collection is clearly innovative in this third scheme. It relies on the "cost recovery principle" to justify an internalization of the restoration costs as operational costs which are passed on to water consumers. Indeed, as opposed to the second scheme that exhibits features of a regional tax allocated to public activities through the district budget [38], [37] (p. 280), this third scheme has the company directly charge the costs of land rehabilitation to water users [37] (p. 272).

2.1.2. Cidanau: Funding by a Private Water Company, Management of Incentives Agreements by Local Stakeholders

The second case study is sited in the Banten province, which is located in the western part of the island of Java (see Map 1). The Cidanau river watershed covers 22,036 hectares and most of the land is privately owned, except for a few plantations that are managed by the parastatal company Perum Perhutani and the 2500 ha Rawa Danau National Reserve in the center. Local residents rely heavily on agricultural development and show interest in using forestry systems for fruit and timber [39]. While the causes of land degradation remain unclear, many acknowledge locally that there has been an increase in illegal farming and migration to the area after the 1998 economic crisis. Both the Rawa Danau National Reserve and the surrounding public forests are affected by this degradation [39].

Land degradation in the Cidanau watershed is thus a source of concern because of soil erosion and surface rainwater runoff. The Rawa Danau swamp area downstream faces eutrophication and sedimentation threats [40,41], and the water quality and average flow of the Cidanau river have decreased [41].

PT Krakatau Tirta Industry (KTI) collects water near the Cidanau river mouth. The water is then processed and distributed to a number of users including (i) the regional public water supply company PDAM (as in the Lombok case) and (ii) another 120 industrial users. While water supplies are currently sufficient to meet the needs of users, KTI staff expressed concerns about the future given that water demand is expected to steadily increase and the above-mentioned environmental problems could lead to a further decrease in water availability and quality, especially during dry seasons. For these reasons the state agency in charge of the watershed management (BPDAS) undertook forest restoration measures in the mid-1990s; in parallel, KTI has been distributing free seedlings to promote reforestation efforts in the watershed.

However, according to concurring views gleaned from interviews with local key informants, such efforts have not met expectations because of poor coordination and unsatisfactory governance. Therefore, in 1998 a broader group of stakeholders established the multi-stakeholder Cidanau Catchment Communication Forum (FKDC) that includes representatives from government agencies (Forest Office, Agriculture Office, provincial and district planning agencies and BPDAS), universities, upstream and downstream farmers, private companies (e.g., KTI), and NGOs (e.g., Rekonvasi Bhumi). In 2002, the Forum received legal recognition with a decree issued by the governor of the Banten province.

The concept of downstream-upstream payments was first introduced to Cidanau stakeholders in 2002 by the German Technical Cooperation (GTZ) and the national NGO LP3ES [42,43]. This move was part of a broader project to develop PES in several watersheds in Indonesia, under the coordination of the International Institute for Environment and Development (IIED). While options were being considered in 2002, a member of the local NGO Rekonvasi Bhumi visited the renowned FONAFIFO Costa Rican PES scheme whereby landowners are paid for sustainably managing their land. During this visit, he understood that conditionality was a real innovation compared to land rehabilitation and reforestation programs in Indonesia, which explains its introduction to the Cidanau scheme. In 2004, the service beneficiary KTI agreed to participate in and fund a PES scheme using the services of FKDC as an intermediary. In principle, KTI would pay the FKDC to support forest management including forest restoration in the watershed, and FKDC would in turn contract with upland farmer groups to plant on their private lands. Payments on both sides—funding and incentive distribution—would be on the condition of satisfactory reports by a monitoring team.

2.2. Methods

This article is an institutional assessment of several PES schemes that are underway in Indonesia. In both case studies, fieldwork was undertaken by two economists and one political scientist during 2012 and 2013. Research techniques included numerous in-depth semi-structured interviews in addition to the analysis of secondary data, from the reports of NGOs and other stakeholders to pieces of legislation and peer-reviewed articles.

At the program level, we interviewed key informants from the main stakeholder institutions: government officials (e.g., the forestry department), intermediary organizations (the essential roles of which are described), companies as main service beneficiaries (water supplier or producer), and NGOs. These interviews led to the collection of data on institutional design and changes, and to the analysis of stakeholder motivations and PES rationale.

At the village level, we interviewed farmer group leaders who were participating in one of the PES schemes under assessment. Their views provided us with relevant first-hand information about the governance of the schemes, their implementation in the field, and their evolution. Data were also collected at the farmer level with focus groups and individual interviews, giving us a comprehensive understanding of farmers' views, the level of information-sharing, and their participation in decision-making processes.

All three levels of observation combine to enable an assessment of the governance structures of two PES in Indonesia, with an analysis of the strategic relationships between the stakeholders involved in these schemes. It provides the framework for the discussion of our research question on institutional change and the effectiveness of new approaches to forest restoration. Collaboration was sought with locally-active research institutions (Bogor Agricultural University, the World Agroforestry Center-ICRAF) and all information collected at different levels could be consistently triangulated with information collected at another level.

3. Results: Institutional Analysis

Results are presented for the two cases and insights are drawn from the data that relate to funding (particularly Lombok) and incentives (particularly Cidanau) so as to assess the situation from both sides of the PES table. This analysis provides the basis for the discussion section where information from both cases is combined to address the research question and the two assumptions.

3.1. Lombok: An Intriguing Process of the Embedment and De-Embedment with Public Policies

3.1.1. Embedment into Public Policies with the Enactment of a Regulation to Secure Funding

The evolution of the Lombok scheme with its three consecutive PES versions is briefly described above. The latter two versions are taking place in parallel, if not in competition with one another. While such an approach may seem complicated at first glance (see Figure 2), an institutional analysis enables a better understanding from a governance point of view.

Figure 2. Governance structure of the three PES in Lombok. Source: [44].

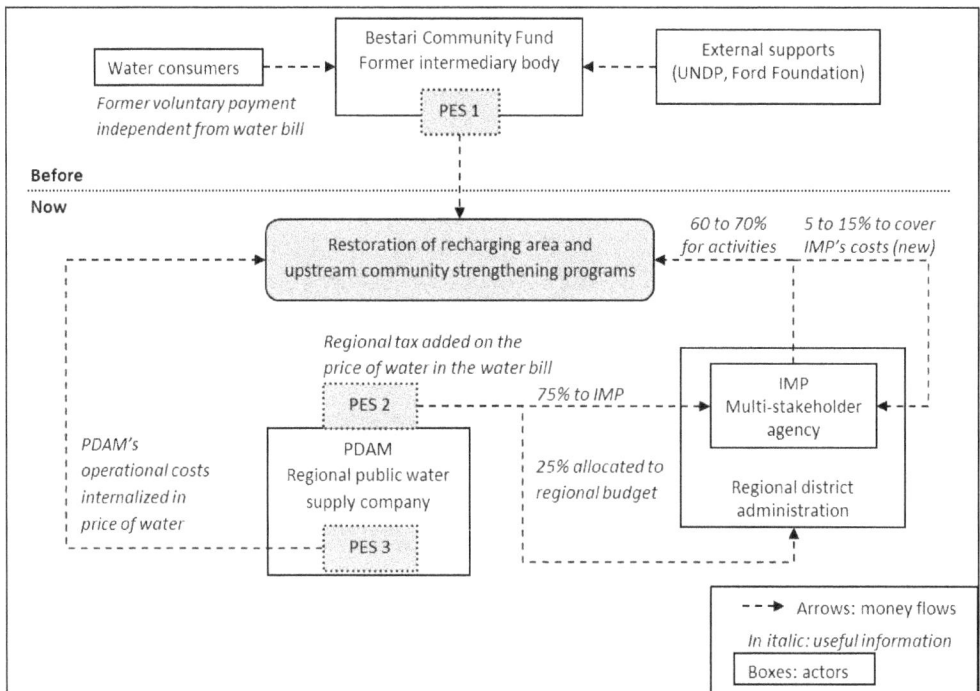

The substitution of PES 1 by PES 2 was clearly justified by the need to secure and enlarge the funding potential of the scheme, so as to increase its capacity to induce forest restoration with the distribution of incentives to farmers. The first PES was initiated by private actors in collaboration with the main service beneficiary, namely PDAM, the public water supply company. In the Lombok context, economic studies of the willingness-to-pay and economic valuations of the environmental services appeared innovative, but the resulting impacts were fated to be anecdotal. The shift from private action to regulation was a consequence of the decision to rely on the PDAM billing system. Even if the willingness-to-pay was high, consumers did not pay spontaneously. It was thus logical to search for another way to collect financial resources.

A legislative process was launched to endorse the new regional tax, as required by national fiscal regulation. It resulted in the establishment of a very limited tax per contributor that is added to monthly bills. The amount depends on the payer: 24,000 households pay USD 0.1 per month (the price of a cigarette), while business entities pay USD 0.2 per month and commercial water producers pay just USD 0.001 per cubic meter of water produced. These new contributions are minimal for each contributor but substantial overall.

Due to the reluctance to pay a new tax in a context where contributors have doubts about the reliability of public authorities to manage the funds, it was decided to establish an intermediary body representing a majority of stakeholders. This multi-stakeholder agency (IMP) involves representatives of civil society and the public sector. This step arguably represented a great move towards participatory governance and enhanced transparency in decision-making, which increased the popularity of the scheme compared to previous land rehabilitation programs in the area.

With the PES 2, based on a political consensus, 75% of collected funds were initially earmarked for PES field activities with the remaining 25% being attributed to the district budget to cover IMP's operational costs. However, recent changes have led to the 75% being used to cover IMP's costs as well, instead of the district budget (see Figure 2), hence reducing the funding available for field activities [37] (p. 270). This move shows less political will than expected in addressing the provision of water services with the PES. Despite a lack of publicly available data on money issues, we were told by IMP that IDR 100 million (USD 10,000) were spent annually in the field, an amount expected to increase in 2013. Over the first two years (2010–2011), 10 agreements materialized with farmer groups, but the contracts did not stipulate conditions on the provision of environmental services once restoration had been performed. Once funding had been received, farmers were free to manage their lands according to their own preferences, and this potentially includes logging the planted trees when mature. Legally, the agreements between IMP and farmer groups are more like legal formalities that are necessary in order to receive a public subsidy that is made on the basis of an administrative decision (unilateral), rather than genuinely negotiated bilateral contracts [37] (p. 271).

This version of the Lombok scheme illustrates the capacity of local actors to engage in up-scaling of funding in order to seek greater impacts with forest restoration. A first and rather naive attempt with voluntary contributions from individual water users led to this refinement, which has characteristics that differ from the original PES concept where funders are free to participate.

3.1.2. A Process of De-Embedment and Cost Internalization… to Enhance Effectiveness?

The main beneficiary from forest restoration activities—the water supply company PDAM—was not entirely satisfied by this course of action and launched a third version of PES in 2011. Indeed, effectiveness was anything but guaranteed with these lax contracts that imposed few, if any, strict conditions on farmers. The third version involves bilateral agreements between the service beneficiary and farmers. However, the district forest administration still intervenes in the management of the scheme on behalf of the PDAM (pre-selection of activities and follow-up of implementation), but contracts and payments directly link the farmer groups with the PDAM. This is an important distinction between PES 2 and PES 3; indeed in PES 2, PDAM is just one of the 16 IMP council members with limited influence on operations, while PDAM is the major actor in PES 3 [37] (p. 204). The second important difference is that conditions are associated with payments in PES 3: tree losses must be replaced at the cost of the farmer. Although payments are made before these conditions are actually verified, our interviews led us to the conclusion that farmers understood that credibility was at stake [37] (p. 276); however the real impacts of this fact have yet to be assessed and cannot be taken at face value.

Our interviews revealed that the high transaction costs of the IMP-led scheme were part of the reason for its replacement, while PDAM sought to lighten the administrative burden, in other words, to reduce bureaucracy. Our interviews also found that PES 3 contracted farmer group leaders thought that procedures were much simpler than with PES 2. The agreement negotiation process is similar in the different PES schemes, but PES 3 follow-up requires less administration for the monitoring, reporting and verification stages [37] (p. 264). A second important point is that PDAM payments are much more generous than IMP ones. In 2011, PDAM disbursed around USD 65,000 (IDR 738 million) to 10 farmer groups, while over the same period IMP distributed around USD 10,000. The larger scale of the PDAM scheme derives from the wider scope of its payment collection process. Indeed, to internalize the costs of water service provision, *i.e.*, PES activities on the ground, the water company collects IDR 1,000 per month (USD 0.01) from all of its 75,000 subscribers in three districts (compared to the 24,000 households subject to the tax in the West Lombok district with PES 2). Farmers prefer to join the PDAM scheme (PES 3) when given the choice, even if the conditions are more restrictive. This competition between the two might explain why IMP is currently trying to move its scheme into new areas where agreements have not yet been signed with the PDAM.

Based on the information collected from key informants, it appears that PDAM had a specific motivation for establishing a new and parallel scheme, specifically, to raise its profile and reputation. With the incorporation of forest restoration costs into the company accounts (PES 3), which are formally included in the water bill as part of water production costs, as opposed to a tax that is imposed by regulation and presented separately on the bill (PES 2), PDAM presents itself as a generous contributor with more attractive contracts for farmers, rather than as a tax collector [37] (p. 272).

This third version consists legally of administrative contracts, a hybrid between a private transaction and a delegation of a public task. When PDAM negotiates a contract, it acts in a similar way to private actors, despite its public legal status. The legitimacy of such payments is based on the contribution provided to the public good and relies on a formal legal basis (although one which is largely ignored) stating that all Indonesian public water supply companies can include restoration activities in their operational costs [37] (p. 170). In contrast, PES 2 consists in the implementation of a public regulation (a *perda*, a regional law, enforced in a *perbup*, a district ordinance). Both are regulated by public law, but they fundamentally differ in nature.

As a conclusion and based on the explanations and details above, it appears that three types of contracts, regulating three distinct types of relationships, are used to conduct similar activities in the same area in a different institutional manner: private contracts in PES 1, implementation of a public regulation in PES 2 and administrative contracts in PES 3 [37] (p. 281). Therefore, from a governance point of view, we observe differences in terms of voluntary (private contracts) or mandatory (public regulation) financial contributions by service beneficiaries, and the role of public authorities in organizing and controlling the transactions (public regulation supervises tax collection with PES 2, but a parastatal company is in charge with PES 3).

3.2. Cidanau: A New Governance without Guarantees of Improved Targeting and Decision-Making

3.2.1. Farmer Groups and the Multi-Stakeholder Agency as Two Key Components of the Governance Structure

This scheme involves two different contracts: the intermediary makes agreements with both the buyer of the service and its provider (see Figure 3). On one side, a Memorandum of Understanding (MOU) was signed in 2005 with the private water producer KTI (renewed in 2010) as the funder, leading to an annual payment of USD 350 per hectare per year for planted and/or conserved forest. Most MOU conditions were decided on by the technical team, which is composed of various stakeholders (KTI, district and provincial planning agencies, the Forest Department and Rekonvasi Bhumi) in consultation with farmers. Building on rules set in previous government land rehabilitation programs, it was decided that a minimum of 25 hectares of *contiguous* lands per farmer group would be necessary for inclusion. Decisions with respect to the number of trees per hectare (which was set at 500) and the level of payment were also inspired by past practice in the national forest rehabilitation program (GERHAN), which was coordinated by the national government [45].

Figure 3. Governance structure of the PES in Cidanau.

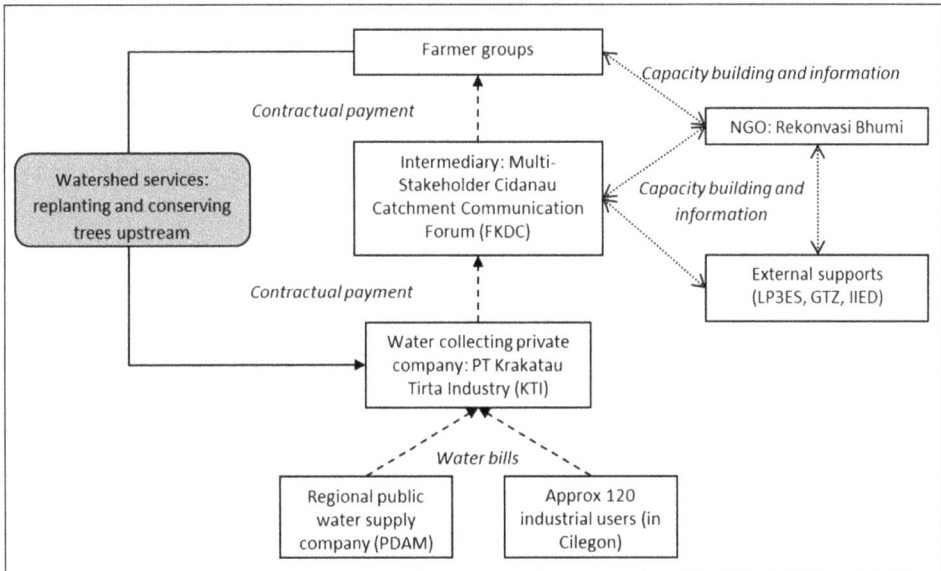

On the other side, contracts were signed between the FKDC and farmer groups as providers for an equivalent period of five years, which included clauses on payment levels and related conditions, including the specification of eligible tree species. The FKDC initially wanted to pay

USD 100 per hectare per year, *i.e.*, a much lower sum than that requested by farmers (USD 250 per hectare per year), but negotiations resulted in a deal being struck at USD 125 per hectare per year (Personal Communication, Pak Hutang, January 10, 2013). Concerning tree species, farmers negotiated for a 70:30 ratio of fruit to timber trees as sufficient for eligibility, contrary to rules commonly followed by past governmental programs.

During the five-year period of the contract, a minimum of 500 trees per hectare must be maintained. The FKDC monitoring team is responsible for ensuring adherence to this stipulation. The team, which includes representatives from a number of stakeholders (e.g., the forest department and KTI), goes into the field once a year to monitor 2.5 hectares of randomly chosen land within each farmer group. Once approval has been given, payments are made to farmer group leaders, who in turn are responsible for the distribution of cash to individual participants. If the team submits a negative report, *i.e.*, if it discovers that even one farmer failed to meet the conditions, then payments are terminated for the whole group. Since the beginning of the scheme's implementation in 2005, two groups breached their contracts, while two others renewed theirs for a further five years, out of a total of eight farmer groups that have been involved at some point.

3.2.2. Business as Usual?

All interviews with key informants confirmed the widespread opinion that the FKDC technical team had a strong tendency to make contracts with farmer groups that it had prior experience working with in various other programs. Its choices were also influenced by the good organizational capacity that these farmer groups had demonstrated in the previous programs. Following on from this, the selection of individual owners and their land remains in the hands of the farmer group leader, so long as those selected meet the requirement of having at least 25 hectares of contiguous land. As a result of this tendency, much land where PES efforts are critically needed, for example land that is steeply sloping, has a high risk of soil erosion or low forest cover, may be excluded from the program; or if it is covered this could be merely coincidence.

The fact that land selection is practically carried out on the basis of social criteria rather than scientific assessment is of critical importance. Indeed, one might question the relevance of PES-funded forest restoration if the provision of environmental services is not high on the agenda, which the analysis of the targeting process suggests. Another article [46] conducted an in-depth investigation into this hypothesis through an extensive survey with more than two-thirds of the scheme's participants (270 interviewees out of 382 participants). The results showed that most of the land engaged in the program already had good forest cover prior to its enrollment, with almost three quarters of participants not requested to plant trees on their lands. Moreover, more than a third of participants described social motivations as the basis for their decision to enroll [46].

3.2.3. Transparency and Decision-Making: Towards Real Innovation?

Qualitative observations and key informant interviews tend to show that participating farmers have a limited understanding of the program and that farmer group leaders retain most of the information. This is a consequence of negotiating and managing contracts with groups as opposed to individuals. As stated by a high-level KTI staff member when asked about cash distribution and internal communication within the farmer groups, "we do not want to look into their local politics" (interview with a KTI Director, Thursday, January 10, 2013). As a result, the amount of knowledge that circulates among participants largely depends on the desire and capacity of group leaders to disseminate information within the group.

We noticed that participants had a good knowledge of the rules in general, although only a few could quote all of them. For instance they were well-aware of the requirement to have more than 500 trees per hectare on their lands to receive funding, but a majority failed to mention the requirement that all lands had to be contiguous over 25 hectares. In fact, it appeared that the role of the farmer group leader was perceived as central, with many respondents declaring to be "actually selected by the farmer group leader". This could mean that local leaders involved in this PES were somehow playing the role of "regulator", whereas these instruments are presented as market-oriented, as opposed to national public programs where public authorities are expected to regulate.

Another critical observation in the field was that participants only had a limited knowledge of the financial amounts that they should receive in the near and mid-term future, and the schedule for these payments, assuming that they met the contractual conditions. This finding was confirmed by [46] who reported that a large majority of households did not know the payment schedule or the amount that they would receive for their next payment. These results point to a lack of transparency and the limited dissemination of information about the PES scheme.

Other observations could also be interpreted as support for the view that the amount of information given to participants is far from satisfactory and the decision-making processes remain opaque. Indeed, the farmer group leader was named by a large majority of participants in response to questions about the persons in charge of determining rules and payments. It is striking that other stakeholders with a strong involvement in contract design were almost completely forgotten: the intermediary FKDC, the water supply company KTI, and representatives from Rekonvasi Bhumi. Moreover, only a handful of participants saw themselves as having a voice in the negotiations about rules and payments, whereas PES are presented as innovative policy instruments that make negotiation and participation a priority.

4. Discussion: Do PES Improve the Governance of Forest Restoration?

Our objective is not to position large-scale governmental programs and PES as opposite ends of a scale of policy instruments for forest restoration; rather we find a continuum of situations in practice. Policy instruments are multi-dimensional: governmental programs can deliver incentives while PES can be designed and implemented by governments. Nonetheless, as a starting point for our analysis, we used the reported weaknesses in terms of governance of traditional public programs for forest restoration in Asia-Pacific [16].

Therefore, instead of comparing two large groups of policy instruments that are artificially separated from one another along the lines of public, traditional and large-scale *versus* private, innovative and local, our study looks at the governance implications of PES through the investigation of two assumptions. The first is that multi-stakeholder agencies as PES intermediaries represent an institutional innovation, positioned between the collection of funds and the distribution of incentives (as opposed to top-down land rehabilitation programs); while the second assumption is that specific individual (or collective) results-oriented contracts with associated conditions attached to payments (as opposed to corporate subsidies or daily salaries) are essential to the success of PES programs.

4.1. First Assumption: PES Intermediaries Represent an Institutional Innovation

Regarding the first assumption, a key governance feature that is present in both Indonesian PES cases is the creation and influential role of a multi-stakeholder agency, which has responsibility for the management of the distribution of incentives among service providers. However, there are striking differences between the two cases. In Lombok, the multi-stakeholder agency was presented as a means to make the tax more palatable to water users in a context where there is mistrust in the government's ability to manage public money. This was the main justification for the creation of the scheme, along with good prospects for a high standard of fund management. However, it appeared that the forest agency benefited from the uneven distribution of power among stakeholders, and was in a position to promote its own priorities using PES financial resources in a context of low budgets allocated to forest agencies. As a consequence, the water distribution company decided to create a parallel scheme that would put environmental services at the center again. By taking this step, the water company no doubt intended to raise its profile and reputation as well as to challenge the power of the forest agency, in addition to addressing other factors such as the high transaction costs.

The non-linear process is the crux of the matter and the most interesting part of the story: early embedment of the PES into public policies with a reliance on regulation to set a specific tax on water users with the creation of the multi-stakeholder agency; followed by a de-embedment, through the creation of a financing mechanism that is fully integrated into the business model of the water supply company. This de-embedment process is expected to strengthen the effectiveness of financial expenditure for the purposes of service provision, or at least address cost-effectiveness issues. Indeed, some observations indicated that fund management by the existing multi-stakeholder agency (PES 2) had weaknesses: the number of contracts finalized so far is limited, and the agency recently decided to allocate to the district budget the share of the collected taxes previously earmarked for covering the implementation costs. It might indicate the temptation of embezzlement that arises in certain contexts when public administrations take the lead, which is precisely the reason why new PES-like experiments are highly praised, as opposed to more traditional governmental programs. Therefore, in this particular case study, the creation of a multi-stakeholder agency might not be a guarantee for better governance.

The situation in Cidanau tells us a different story; here the multi-stakeholder agency remains the principal and widely recognized actor in the area. The agency is also seemingly dominated by one stakeholder from civil society which has a great influence owing to its past accomplishments. Yet another important layer exists at the interface between the agency and individual farmers, namely the farmer group leaders, and it was this layer that was a focus of our study in Cidanau. Our field observations showed that these farmer group leaders played a vital role in the scheme, a finding that was confirmed by the two instances of breach of contract, both of which could have been avoided with appropriate action on their side. The problem is that there is much variability in the management abilities among the farmer group leaders. Governance in the Cidanau situation depends a great deal on the capacities of these farmer group leaders, and the intermediary agency neither guarantees good governance nor has a negative impact in this regard. On the whole, the internal governance of the farmer groups appears to be decisive for the sustainable effectiveness of forest restoration efforts.

Another key observation is the inability of this governance structure to ensure the satisfactory targeting of lands for restoration. Having a multi-stakeholder set up provides no guarantee that participants will be identified and selected in a neutral way and that decisions will be based only on scientific information with regard to the provision of environmental services. Social connections were favored as a criterion for farmer enrollment (and hence land selection), which in our opinion constitutes a weakness of the scheme as it puts effectiveness at risk. In other words, land with the highest potential contribution to environmental services provision is probably not more likely than other land to be earmarked for forest restoration. This result is consistent with other empirical cases of small-scale watershed projects. In Central America, it was demonstrated that the choice of PES participants results from a complex social process rather than a rational technical assessment [47]. These authors conclude that payments only provide complementary "support" for activities that farmers would have carried out for social and cultural reasons. In Peru and Ecuador, it was contended that better spatial targeting could be achieved in two watersheds in order to include genuinely critical areas [48].

At a larger scale, our finding also complements the aforementioned observation that large-scale governmental forest restoration programs in Asia-Pacific have sometimes resulted in forest conversion prior to planting [16], which is another hazardous method of land targeting from the perspective of forest restoration.

4.2. Second Assumption: Results-Oriented Contracts Are an Essential Aspect of PES

Regarding the second assumption under investigation, the results-oriented conditions that constitute a key feature of PES as a new approach to forest restoration are not particularly strong. While their full impact remains to be demonstrated, the two case studies examined here provide lessons that differ from our assumption. In Lombok, few (if any) PES 2 conditions are enforced, and it is not yet clear whether PES 3 will be any better at putting pressure on farmers to carry out effective land-use changes. Besides which, the contracts are at an early stage and cannot compete with larger scale intensive reforestation programs financed by regional and provincial forest administrations. That said, the three successive versions of the scheme are assumed to have the potential to eventually tackle causes of deforestation owing to their capacity to change local perceptions and habits. They rely on the active participation of farmers to make proposals and are not perceived as top-down public policies; as a consequence, they are thought to have an indirect leverage effect that may exceed the direct corrective effort of more "traditional" restoration programs. The latter usually involves the payment of salaries to local laborers who are hired to plant trees but have little stake in their maintenance in subsequent years.

In Cidanau, these conditions are more stringent, which is demonstrated by the fact that infringements led to the breaching of two contracts. The credibility of the threat to withhold payments is also a central element of PES governance and one that is seen as a step towards greater effectiveness compared to traditional governmental programs because it generates better results than salaries paid to locals in return for daily labor, or the opaque distribution of subsidies to well-connected corporations. In this regard, despite many examples of individuals having a poor understanding of these conditions and their implications for future payments, we could indeed observe a certain level of achievement. Yet we also observed a tendency to enroll farmers who might not have dramatically changed their business-as-usual activities, which means limited additionality and a low level of threat with the conditions. In addition, farmer group leaders have a certain amount of latitude to prevent the breaching of contracts when conditions are not met.

Overall, the two sites exhibit the same characteristic that is detrimental to effectiveness: most stakeholders have a vested interest in the perpetuation of the scheme, whatever its level of success in terms of sustaining the provision of environmental services. In other words, NGOs, local authorities, research institutions, and even private companies—as service beneficiaries when they use funds from Corporate Social Responsibility budgets—prefer to avoid apparent failure at any cost. In practical terms, failure is understood as the cessation of payments rather than a lack of service provision, which is clearly a controversial view. The problem is that, regardless of the degree of stringency for conditions, effectiveness is eliminated whenever additionality is absent or the targeting of service providers is irrelevant. Therefore, a "winning" strategy (for a number of stakeholders but certainly not from an environmental point of view) would be for payers and intermediaries to demonstrate that strong conditions are attached to sustained payments, while at the same time involving the most easily targetable service providers. This typically implies that farmers do not attempt to change their activities and there is no guarantee that the right farmers are brought on board.

5. Conclusions

This article discusses the capacity of innovative policy instruments such as PES to improve the governance of forest restoration activities compared to more traditional large-scale governmental programs. To do so, two assumptions were investigated, the first regarding the establishment of multi-stakeholder agencies as intermediaries and fund managers; and the other concerning the inclusion of conditions in the contracts with service providers. Both of these assumptions are believed to enhance forest restoration efforts.

An initial finding was that intermediary bodies are certainly not sufficient to guarantee success. As shown in different ways by the two cases under investigation, outcomes were greatly dependent on the internal governance of these bodies. While virtually all local stakeholders were represented, in each case we found that about one was able to dominate the decision-making process: the forest agency in Lombok and a local NGO in Cidanau. Interestingly, the main service beneficiaries in each case study adopted opposite strategies in reaction to this domination by another actor: the public water company in Lombok moved on and created its own scheme, whereas the private water company in Cidanau decided to keep the ball rolling, its expectations being little more than the nurturing of its image. The situation in Cidanau might however deserve a more positive appraisal given that the local NGO involved understands the difficulty in achieving a high degree of effectiveness but is making incremental changes towards improvement. For instance, the somewhat shaky governance of many of the farmer groups is identified as one area of reform for the future. Reforms are probably more difficult to undertake in Lombok where the intermediary body is de facto controlled by local administrations. It remains to be seen whether stakeholders can improve the scheme based on its existing format, instead of creating an alternative, as PDAM has done.

A second conclusion is that even when conditions exist, they do not guarantee success. Not only because they can be applied to the wrong participants in the sense that their business-as-usual activities remain unchanged, but also because there is a common interest among many stakeholders to keep the schemes alive and visible. Since the service beneficiaries do not have any alternative options, they must find ways to ensure that forest restoration takes place on the ground, even if it means ignoring (temporary) failures when the wrong plots are targeted and there is no additionality. In this context, conditions can be seen as a means to raise awareness among service providers and to increase the chances of success in future rounds. Another interpretation would be that conditions are designed in response to local capacities and not the other way around; in other words these conditions would encourage rather than strictly regulate service providers.

Although our results reveal the limited effectiveness of the schemes that aim at promoting forest restoration despite innovations in their governance owing to PES schemes, either because the scale is too small, additionality is not proven or targeting is flawed, our overall conclusion is that local stakeholders have a great ability to adapt and make progress. In both case studies, processes were initiated by international actors eager to replicate the PES model as conceptualized in foreign institutions: the London-based IIED coordinated the project in Cidanau in the early stages, and international organizations such as the Ford Foundation, USAID and UNDP were influential at the very beginning of the process in Lombok. Yet directions have largely diverged over time, and it is undeniable that a sense of ownership has developed among local stakeholders. While one case exhibited a very dynamic evolution with three successive versions of PES and an unstable reliance on regulation and public policies (Lombok), the other example has proven to be more resilient in design with a classical "private beneficiary-intermediary-land users" set up (Cidanau). This finding is interesting because both schemes were influenced by the international discourse advocating new ways to foster good forest management, and both schemes addressed the same water services in a same country. Therefore, having such diversity in terms of governance is a key issue: rules, modalities of intermediation and participation, fund collection, conditions, and payments, are all elements that differed in order to adapt to the local context.

Ultimately, and despite the limited scale of forest restoration activities and a lack of evidence for the effectiveness of these PES schemes with respect to service provision, we find optimism in the future possibilities for these new ways to govern forest restoration in a developing country context. Lessons from past failures in governmental programs—or at least assumed failures—are in the minds of local proponents of innovations in governance for forest restoration initiatives. Innovations can deliver and yield positive results, despite resistance from local administrations or state agencies that are used to taking advantage of opportunities for embezzlement and thus want these opportunities to continue. Yet these public actors will remain indispensable for the provision of these public goods, and it might prove to be more productive to find enabling conditions for their positive participation, rather than just trying to bypass them.

Acknowledgments

This research was partly funded by the ERA-Net BiodivERsA, with the French national funder Agence Nationale de la Recherche (Convention 2011-EBID-003-01), part of the 2011 BiodivERsA call for research proposals.

The field research in Lombok has been financed by a fellowship granted by the Swiss National Science Foundation (SNF) (Reference PBLAP1_140045).

Conflicts of Interest

The authors declare no conflict of interest.

References and Notes

1. FAO. *Global Forest Resources Assessment 2010: Main Report*; FAO Forestry Paper; Food and Agriculture Organization: Rome, Italy, 2010; Volume 163.
2. TEEB. *The Economics of Ecosystems and Biodiversity for National and International Policy Makers—Summary: Responding to the Value of Nature*; United Nations Environment Programme (UNEP), The Economics of Ecosystems & Biodiversity: Geneva, Switzerland, 2009.
3. Sterner, T.; Coria, J. *Policy Instruments for Environmental and Natural Resource Management*, 2nd ed.; RFF Press: Washington, DC, USA, 2011.
4. We are aware of the discussions around the use of terms such as environmental, ecological or ecosystem services, but in this article we prefer to use only one of these terms consistently, assuming that these distinctions are not relevant for the purpose of our analysis.
5. Armsworth, P.R.; Chan, K.M.A.; Daily, G.C.; Ehrlich, P.R.; Kremen, C.; Ricketts, T.H.; Sanjayan, M.A. Ecosystem-service science and the way forward for conservation (editorial). *Conserv. Biol.* **2007**, *21*, 1383–1384.
6. Engel, S.; Pagiola, S.; Wunder, S. Designing payments for environmental services in theory and practice: An overview of the issues. *Ecol. Econ.* **2008**, *65*, 663–674.
7. Forest restoration is defined in this special issue as the process to assist the recovery of damaged forest ecosystems. Although the cases studied in this article were not designed with this definition of forest restoration in the minds of their promoters, but rather as attempts to fund reforestation with a diversity of species, we argue that the implications for governance would be identical. Therefore, PES can be viewed as vehicles of forest restoration as long as ecosystem services are targeted.
8. Chen, X.; Lupi, F.; Viña, A.; He, G.; Liu, J. Using cost-effective targeting to enhance the efficiency of conservation investments in payments for ecosystem services. *Conserv. Biol.* **2010**, *24*, 1469–1478.

9. Muñoz-Piña, C.; Guevara, A.; Torres, J.-M.; Brana, J. Paying for the hydrological services of Mexico's forests: Analysis, negotiations and results. *Ecol. Econ.* **2008**, *65*, 725–736.

10. Leimona, B.; Joshi, L.; van Noordjwijk, M. Can rewards for environmental services benefit the poor? Lessons from Asia. *Int. J. Commons* **2009**, *3*, 82–107.

11. Corbera, E.; Kosoy, N.; Martinez Tuna, M. Equity implications of marketing ecosystem services in protected areas and rural communities: Cases from Meso-America. *Glob. Environ. Chang.* **2007**, *17*, 365–380.

12. Pirard, R. Payments for Environmental Services (PES) in the public policy landscape: "Mandatory" spices in the Indonesian recipe. *For. Policy Econ.* **2012**, *18*, 23–29.

13. Fisher, J. No pay no care? A case study exploring motivations for participation in payments for ecosystem services in Uganda. *Oryx* **2012**, *46*, 45–54.

14. Garcia-Amado, L.R.; Ruis Perez, M.; Barrasa Garcia, S. Motivation for conversation: Assessing integrated conservation and development projects and payments for environmental services in La Sepultura Biosphere Reserve, Chiapas, Mexico. *Ecol. Econ.* **2013**, *89*, 92–100.

15. Vatn, A. An institutional analysis of payments for environmental services. *Ecol. Econ.* **2009**, *69*, 1245–1252.

16. Barr, C.; Sayer, J. The political economy of reforestation and forest restoration in Asia-Pacific: Critical issues for REDD+. *Biol. Conserv.* **2012**, *154*, 9–19.

17. Sierra, R.; Russman, E. On the efficiency of environmental service payments: A forest conservation assessment in the Osa Peninsula, Costa Rica. *Ecol. Econ.* **2006**, *59*, 131–141.

18. Mauerhofer, V.; Hubacek, K.; Coleby, A. From polluter pays to provider gets: Distribution of rights and costs under payments for ecosystem services. *Ecol. Soc.* **2013**, *18*, 41.

19. Montagnini, F.; Finney, C. Payments for environmental services in Latin America as a tool for restoration and rural development. *AMBIO* **2011**, *40*, 285–297.

20. Le, H.D.; Smith, C.; Herbohn, J. What drives the success of reforestation projects in tropical developing countries? The case of the Philippines. *Glob. Environ. Chang.* **2014**, *24*, 334–348.

21. Bullock, J.M.; Aronson, J.; Newton, A.C.; Pywell, R.F.; Rey-Benayas, J.M. Restoration of ecosystem services and biodiversity: Conflicts and opportunities. *Trends Ecol. Evol.* **2011**, *26*, 541–549.

22. Brancalion, P.H.S.; Viani, R.A.G.; Strassburg, B.B.N.; Rodrigues, R.R. Finding the money for tropical forest restoration. *Unasylva* **2012**, *239*, 15–34.

23. Ciccarese, L.; Mattsson, A.; Pettenella, D. Ecosystem services from forest restoration: Thinking ahead. *New For.* **2012**, *43*, 543–560.

24. Fletcher, R.; Breitling, J. Market mechanism or subsidy in disguise? Governing payment for environmental services in Costa Rica. *Geoforum* **2012**, *43*, 402–411.

25. Lapeyre, R.; Pirard, R. *Payments for Environmental Services and Market-based Instruments: Next of Kin or False Friends?* IDDRI Working Paper; Institute for Sustainable Development and International Relations: Paris, France, 2013.

26. Wunder, S. Of PES and related animals. *Oryx* **2012**, *46*, 1–2.

27. Broughton, E.; Pirard, R. *What's in a Name? Market-based Instruments for Biodiversity*; IDDRI Analyses; Institute for Sustainable Development and International Relations: Paris, France, 2011.

28. Calder, I. Forests and Hydrological Services: Reconciling public and science perceptions. *Land Use Water Resour. Res.* **2002**, *2*, 2.1–2.12.

29. FAO. *Forests and Water*; FAO Forestry Paper 155; Food and Agriculture Organization: Rome, Italy, 2008.

30. Van Dijk, A.; Keenan, R. Planted forests and water in perspective. *For. Ecol. Manag.* **2007**, *251*, 1–9.

31. Bruijnzeel, L.A. Hydrological functions of tropical forests: Not seeing the soil for the trees? *Agric. Ecosyst. Environ.* **2004**, *104*, 185–228.

32. Dye, P.; Versfeld, D. Managing the hydrological impacts of South African plantation forests: An overview. *For. Ecol. Manag.* **2007**, *251*, 121–128.

33. Keenan, R.; van Dijk, A. Planted Forests and Water. In *Ecosystem Goods and Services from Plantation Forests*; Bauhus, J., van der Meer, P., Kanninen, M., Eds.; Earthscan: London, UK; Washington, DC, USA, 2010; pp. 77–95.

34. Holl, K.D.; Aide, T.M. When and where to actively restore ecosystems? *For. Ecol. Manag.* **2011**, *261*, 1558–1563.

35. Chandler, D.G. Reversibility of forest conversion impacts on water budgets in tropical karst terrain. *For. Ecol. Manag.* **2006**, *224*, 95–103.

36. Nugraha, P. Number of Natural Springs in West Nusa Tenggara Sees Sharp Fall in Recent Years. *The Jakarta Post*, 29 May 2011.

37. De Buren, G. *La régulation des interdépendances entre les forêts et l'eau domestique en Indonésie: études de cas sur le site de Lombok*; Idheap Working Paper; Swiss Graduate School of Public Administration: Lausanne, Switzerland, 2013. Available online: http://idheap.ch/deBuren2013Lombok (accessed on 10 March 2014).

38. The tax created by the second PES is collected through monthly water bills, which adds to the confusion between both funding mechanisms. While it is not part of the water production costs, neither does it constitute a source of income for the water supply company. It is thus not a process of internalization on behalf of the company, and was even declared illegal by a commission of the national Financial Advisory Board (BPKP).

39. Yoshino, K.; Ishikawa, M.; Setiwawn, B.I. Socio-economic causes of recent environmental changes in Cidanau watershed, west Java, Indonesia: Effects of Economic Crises in Southeast Asia in 1997–1998 on Regional Environment. *Rural Environ. Eng.* **2003**, *44*, 27–41.

40. Adi, S. *Proposed Soil and Water Conservation Strategies for Lake Rawa Danau, West Java, Indonesia*; Water Resources System, Hydrological Risk, Management and Development No. 281; International Association of Hydrological Sciences Publication (IAHS): Wallingford, UK, 2003.

41. Yoshino, K.; Ishioka, Y. Guidelines for soil conservation towards integrated basin management for sustainable development: A new approach based on the assessment of soil loss risk using remote sensing and GIS. *Paddy Water Environ.* **2005**, *3*, 235–247.

42. Munawir, S.; Vermeulen, S. *Fair Deals for Watershed Services in Indonesia: IIED Natural Resource Issues*; International Institute for Environment and Development: London, UK, 2009; Volume 9.
43. This NGO was also involved in the discussions in Lombok as part of a project with IIED to promote PES in the country. This can be viewed as a factor of standardization, but our analysis also points to great differences in terms of design and evolution between both sites.
44. Pirard, R.; de Buren, G. *Payments for Watershed Services in Indonesia (Lombok): Uncovering Actor's Strategies in a "Success" Story. Factsheet for the Multi-Stakeholder Dialogue (September 13 2013)*; Food and Agriculture Organization: Rome, Italy, 2013.
45. Leimona, B.; Pasha, R.; Rahadian, N.R. The Livelihood Impacts of Incentive Payments for Watershed Management in West Java, Indonesia. In *Livelihoods in the REDD? Payments for Environmental Services, Forest Conservation and Climate Change*; Tacconi, L., Mahanty, S., Suich, H., Eds.; Edward Elgar: Cheltenham, UK, 2010; pp. 106–129.
46. Lapeyre, R.; Pirard, R. Payments for environmental services in Indonesia: What if economic signals were lost in translation. *Ecol. Econ.* **2014**, submitted.
47. Kosoy, N.; Martinez-Tuna, M.; Muradian, R.; Martinez-Alier, J. Payments for environmental services in watersheds: Insights from a comparative study of three cases in Central America. *Ecol. Econ.* **2007**, *61*, 446–455.
48. Quintero, M.; Wunder, S.; Estrada, R.D. For services rendered? Modeling hydrology and livelihoods in Andean payments for environmental services schemes. *For. Ecol. Manag.* **2009**, *258*, 1871–1880.

From Target to Implementation: Perspectives for the International Governance of Forest Landscape Restoration

Till Pistorius and Horst Freiberg

Abstract: Continuing depletion of forest resources, particularly in tropical developing countries, has turned vast areas of intact ecosystems into urbanized and agricultural lands. The degree of degradation varies, but in most cases, the ecosystem functions and the ability to provide a variety of ecosystem services are severely impaired. In addition to many other challenges, successful forest restoration of these lands requires considerable resources and funding, but the ecological, economic and social benefits have the potential to outweigh the investment. As a consequence, at the international policy level, restoration is seen as a field of land use activities that provides significant contributions to simultaneously achieving different environmental and social policy objectives. Accordingly, different policy processes at the international policy level have made ecological landscape restoration a global priority, in particular the Convention on Biological Diversity with the Aichi Target 15 agreed upon in 2010, which aims at restoring 15% of all degraded land areas by 2020. While such ambitious policy targets are important for recognizing and agreeing upon solutions for environmental problems, they are unlikely to be further substantiated or governed. The objective of this paper is thus to develop a complementary governance approach to the top-down implementation of the Aichi target. Drawing on collaborative and network governance theories, we discuss the potential of a collaborative networked governance approach and perspectives for overcoming the inherent challenges facing a rapid large-scale restoration of degraded lands.

Reprinted from *Forests*. Cite as: Pistorius, T.; Freiberg, H. From Target to Implementation: Perspectives for the International Governance of Forest Landscape Restoration. *Forests* **2014**, 5, 482-497.

1. Introduction

The depletion and conversion of forests and forested lands has turned vast areas of intact ecosystems into degraded landscapes: the Global Partnership on Forest Landscape Restoration (GPFLR) identified across all continents a total area of one to two billion hectares of converted and degraded forest lands [1]. Degradation is the result of land uses, such as unsustainable logging practices, encroachment and overexploitation, or direct and indirect land use changes, in particular for agro-industrial development and urbanization [2]. These human activities are the main causes of terrestrial biodiversity loss; they impair and disrupt the functionality of ecosystems, with mostly negative consequences for the provision of vital ecosystem services at global, regional and local levels [3,4]. Since it depends on the purpose and the perspective of those who assess the state of an ecosystem, there are more than 50 different definitions related to degradation [5]; however, despite significant differences, they all refer to a reduction of the capacity of a forest to provide ecosystem goods and services [6]. With this, degradation and its negative consequences affect present, as well as future generations across the globe, but most specifically, those who directly depend on the services provided by local ecosystems [7]. In this paper, we focus on one specific cross-cutting issue that aims at reversing these trends and their negative consequences: the restoration of degraded forest ecosystems [8].

In 2010, the Parties to the Convention on Biological Diversity (CBD) agreed in Nagoya on its strategic plan, the so-called Aichi targets. Here, ecosystem restoration is a crucial element of the goal "to enhance the benefits to all from biodiversity and ecosystem services". In particular, Aichi Target 15 highlights the above-mentioned synergies between climate change, biodiversity and desertification, while it allows for its quantification): "by 2020, ecosystem resilience and the contribution of biodiversity to carbon stocks has been enhanced, through conservation and restoration, including restoration of at least 15% of degraded ecosystems, thereby contributing to climate change mitigation and adaptation and to combating desertification [9]". Although the target refers to all ecosystems, restoration of forests will be the main focus, given that a major proportion of the identified degraded areas are, or were, forested prior to their transformation. Restoration of degraded lands, especially those once covered by forests, is considered by scientists, non-governmental organizations (NGOs) and other actors as a field of activities that helps to maintain and provide a number of social and environmental services [10,11]. Due to its positive contributions to the sequestration of carbon dioxide and the so-called co-benefits, restoration also plays an increasing role in the negotiations under the United Nations Framework Convention On Climate Change (UNFCCC) on REDD+ [12], an international financing mechanism intended to compensate developing countries that succeed in mitigating land use and forest sector emissions [13].

At most international environmental conferences of the UNFCCC and the CBD, there is agreement that activities with apparent potentials to enhance synergies among the globally agreed upon environmental and social objectives should be promoted. During the 1980s and 1990s, political theories on international relations assumed that implementation of policy targets at state-centered, international regimes, such as the Rio conventions, would occur automatically, trickling-down to local policy levels [14]. However, despite the expressed consensus on ambitious policy objectives and targets, the unabated trends of land use and conversion during the last two decades show that implementation and concrete actions on the ground lag behind, and the problems remain unsolved. Examples are the unabated increase of anthropogenic greenhouse gas emissions, the high rates of ecosystem degradation and conversion [15] and the continuing loss of biodiversity [16,17].

The failure to substantiate the agreed upon policy objectives brings the effectiveness and legitimacy of multilateral environmental agreements (MEAs) into question [18] and has prompted scholarly debates on more effective alternative modes of governance. In our reflections, we focus on the international policy level and the major challenge of resource mobilization, but we acknowledge that there are many other political and technical hurdles associated with the implementation of large-scale restoration at the national and at the local levels. Our main assumption is that without corresponding options for financing such activities, this target cannot be met and that the mobilization of new and additional funding to the levels outlined below requires a well-coordinated and institutional approach: globally and starting from 2013, Aichi Target 15 of the CBD strategic plan implies the necessity to restore annually an area equal to the size of the state of Nepal. A study estimating the costs of complying with Aichi Target 15 has analyzed very heterogeneous examples of restoration activities and estimates that the costs for restoration activities lie between US$500 to 1500 per ha, which is equal to a financing need of US$75 billion by 2020, or more than US$10 billion per year [19]. Another estimate predicts that the funding needed to implement Aichi Target 15 will amount to US$47.6 billion by 2020 [20]. In light of the total amount of global funding currently available for conservation, such investments exceed the capacities of governments by far, especially those of developing countries, where the largest potential for restoration is found. For comparison, the total amount of non-market funding for biodiversity conservation in developing countries is estimated to range between US$13 and 16 billion per year [21]. These figures explain why most debates on the targets of the CBD, the UNFCCC and other conventions are intricately linked with those on the mobilization of corresponding funding sources.

The aim of this paper is to provide perspectives on complementary governance approaches for an effective implementation of Aichi Target 15, in particular on a networked approach for the mobilization of resources through private-public partnerships (PPPs). For this purpose, we first provide an overview on what political scientists refer to as an "institutional landscape"—the main international institutions whose work is directly relevant to this policy objective. We then review literature on international relations and environmental governance theories to draw conclusions about elements and aspects of governance approaches suitable for aligning the different efforts and activities of the many public and non-public institutions working on forest restoration. Methodologically, we base our findings and opinions on desk work and an extensive review of the academic literature on collaborative governance. This is complemented by insights from participatory observation at a plethora of land-use related policy events: Conferences of the Parties of UNFCCC and CBD since 2006, the Bonn Challenge [22] in 2011 (described below) and side events at meetings, such as the forest/landscape days, organized by the Center for International Forestry Research [23].

2. International Public and Non-Public Institutions Promoting Restoration

In this section, we illustrate the continuously increasing number of public and non-public institutions whose work relates to the restoration of degraded lands and whose objectives are overlapping. In particular, we consider a specific issue—a phenomenon that is described by political science scholars as a "fragmented institutional landscape" [24]. In this way, we provide the basis for answering the main question of this paper: how can the work and activities of these many different institutions with overlapping objectives be aligned effectively in a complementary governance approach to overcome challenges related to resource mobilization and the implementation of globally agreed upon environmental policy objectives?

Given the trans-boundary effects of environmental degradation, many political efforts have been made to address the continuing trends and consequences of depleting natural resources, especially of global deforestation and unsustainable land uses. As outlined below, the corresponding debates at the international policy level have led to the establishment of many public institutions during the last four decades since the 1972 United Nations Conference on the Human Environment in

Stockholm. At this milestone of global environmental politics, governments first recognized the link between the quality of the environment and economic development and established the United Nations Environmental Program (UNEP). Fifteen years later, the Brundtland Commission cemented this link in the globally accepted definition of sustainable development, which, until today, represents the common basis for the many institutions that have since been created to deal with environmental issues; in particular, the prominent MEAs agreed upon in 1992: the UNFCCC, the CBD under the institutional roof of UNEP and the United Nations Convention on Combatting Desertification (UNCCD). Less prominent, state-centered policy processes are the United Nations Forum on Forests, the International Tropical Timber Organization and regional processes, such as the Forest Europe Process or the Ramsar Convention on Wetlands. Next to these government-driven institutions, a large number of intergovernmental institutions work on topics directly related to land use, degradation and restoration, e.g., the Food and Agriculture Organization, UNEP and the United Nations Development Program.

While these examples refer only to public institutions, there is also a plethora of non-public institutions, which are active in the same fields, which contribute to the implementation of policy objectives, as well as influence the state-driven processes and which often form networks and partnerships for achieving shared objectives [25]. In the following, we briefly describe some non-state international institutions that are most directly related to Aichi Target 15, as a result of their global involvement in forest restoration programs, but acknowledge that there are many more that would also warrant being listed.

The International Union for the Conservation of Nature (IUCN) is a network that links more than 900 NGOs and 200 public institutions that associate themselves with conservation of the environment. It is represented in most countries of the world and has the status of an official observer organization at the United Nations General Assembly and many international processes. The IUCN assembles and brokers knowledge and best practices through databases, as well as numerous scientific and science-based publications, and it exerts influence on the negotiations of the international environmental conventions mentioned above. In addition, it facilitates hundreds of conservation, restoration and development projects across the globe. One member of IUCN is the Society for Ecological Restoration (SER), another global network with members in more than 70 countries, which is dedicated to the science and practice of "reversing degradation and restoring the Earth's ecological balance for the benefit of humans and nature". On the global level, the SER has established partnerships with international political processes and regularly provides input to the CBD, the Ramsar Convention and the UNCCD. Furthermore, it is linked to other networks, such as Parks Canada and the Wildlands Network, and has established its own online networks (the Global Restoration Network, the Indigenous Peoples' Restoration Network and the Community Restoration Network). Through these network activities, the SER bundles existing competencies and provides the knowledge brokerage necessary for the practical implementation of restoration activities and corresponding policy development at different levels.

The Global Partnership on Forest Landscape Restoration (GPFLR) is another network, initiated by IUCN, the World Wide Fund for Nature (WWF) and the Forestry Commission of Great Britain. Guided by ten principles [26], it pursues the aim "to weave a thread through existing activities, projects, processes and institutions to encourage and reinforce the positive roles and contributions of each of them". The partners are comprised of public actors (donor and beneficiary governments, the secretariats of relevant international policy processes), as well as non-state actors, especially NGOs and renowned research organizations. It catalyzes support for restoration activities at international, national and regional policy levels. Furthermore, it has established a learning network for knowledge brokerage and implementation tools, e.g., the so-called map of opportunity that quantified in a geo-referential map the global potential for restoration and identified main areas of opportunity. Currently, it is being further developed with the aim to refine this global analysis to the national level by combining multiple sources of data, with Mexico and Ghana as pilot countries. Such national assessments allow policy makers, land managers and potential investors to identify relevant local stakeholders for their participation. A similar approach is pursued by the Economics of Ecosystems and Biodiversity (TEEB) Initiative: in 2007, the G8+5 governments agreed to analyze the global economic benefits of biological diversity and the economic costs of its loss. Following this agreement, the German Federal Ministry for the Environment (Bundesministerium für Umwelt, Naturschutz und Reaktorsicherheit, BMUB) and the European Commission (EC) initiated a global study that has resulted in a series of study reports. In the meantime, it has grown into a strong network at the science-policy interface hosted by UNEP, which coordinates national TEEB activities.

In 2011, the BMUB organized, in collaboration with IUCN and the GPFLR, the "Bonn Challenge", a forum for different stakeholders and forest restoration experts (senior officials of national governments and representatives of the Rio conventions' secretariats, scientists, NGOs and business representatives). The objective was to contribute, through concrete actions and pledges, to the implementation of Aichi Target 15 and the REDD+ mechanism negotiated under the UNFCCC. IUCN and the company, Airbus, launched at the Bonn Challenge their "plant-a-pledge" campaign, where governments, business representatives and private people are requested and given the opportunity to make concrete pledges. During the Bonn Challenge and in its aftermath, more than 20 million ha have been pledged to date (by Rwanda, USA, Brazil, Costa Rica and El Salvador). Another 30 million ha of pledges still have to be confirmed (India, the Meso-American Alliance of Peoples and Forests) and more countries are expected to follow. In addition, the senate of the German Economy—a business network of large and medium German enterprises—announced during the event its world-forest-climate initiative, which pursues the objective of finding investors and raising significant amounts of private funding for forest restoration. In addition to these remarkable pledges, the Bonn Challenge has since been mentioned at various high-level political meetings, such as the CBD COP11 in India and the Rio Summit in 2012, where the government of Brazil provided the opportunity for civil society to "vote for the future we want", and the "Bonn Challenge" goal of restoring 150 million ha by 2020 was only topped by the demand for concrete steps to end fossil fuel subsidies.

As a consequence, collaborative governance approaches, such as public policy networks and partnerships between private companies, governmental bodies and civil society organizations, have rapidly gained momentum since the Earth summit in Johannesburg in 2002 [27,28]. In fragmented institutional landscapes, partnerships can tie together different actors with individual rationales "though a complex web of interdependencies in which collaboration is required to achieve individual and common purposes" [27,29]. This relatively new form of public management of private-public partnerships, or type-2 partnerships, is considered a legitimate alternative approach for poorly implemented intergovernmental agreements [30,31].

3. Theoretical Considerations on Environmental and Networked Governance

During the 1980s and 1990s, the literature on international relations focused on hierarchical, state-led policy processes of MEAs, in particular on the above-mentioned Rio conventions, which were expected to effectively respond to global environmental problems [32]. In this view, scholars consider governments to be the decisive actors in top-down processes, and they seldom attribute a role to non-state actors and institutions that extends beyond "observation" [14]. Indeed, the important watch-dog function of observers does not directly shape the policy outcome; however, their significant indirect impacts and their role in raising public awareness and, consequently, the expectations placed on the negotiating governments is widely recognized [33]. In contrast to the aforementioned hierarchical perspective, modern theories on environmental governance attribute a much more important role to non-state actors and take the view that they can (and should) contribute much more than just ensuring the transparency of governmental behavior in negotiations. In particular, they expect non-state actors to contribute to the legitimacy and accountability of policies and their implementation [34]. The shift in these scholarly debates and related research towards less hierarchical and more inclusive political thinking in global governance was spurred by the fact that during the last decade, high public expectations invested in the outcomes of MEAs were repeatedly disappointed, because governments succeeded, at best, in agreeing on ambitious road maps and policy targets, such as Aichi Target 15. These are important, provided they are accompanied by corresponding initiatives and activities for their implementation.

While hierarchical modes and markets have failed as approaches for environmental governance, a large number of alternative governance modes have gained momentum [33,34], stretching from classical state-driven initiatives over PPPs, to purely private, market-oriented mechanisms, such as certification schemes [35]. In contrast to hierarchical, top-down processes, such as the aforementioned MEAs, these approaches are characterized by reciprocal communication and mutual influence between public and non-public actors [27]. In the following, we describe the elements of these networked governance approaches in order to discuss how they can be aligned with the policy targets of MEAs.

3.1. Partnerships and Collaborative "Networked" Governance

PPPs as organizational structures can be distinguished from networks as a governance mode [29]. Naturally, the many emerging partnerships in the context of global policy-making differ considerably regarding their objectives, structures and character. However, they share the common and distinctive feature of pursuing the implementation of public policy objectives through non-hierarchical transnational network structures, which integrate different actors "within a horizontal structure" [33,36]. Collaborative governance approaches that bundle private and public actors in PPPs are seen as

28

having the potential to generate "outcomes that could not normally be achieved by individual organizational participants acting independently" [37]. PPPs undergo a cyclic development "in which different modes of governance assume a particular importance at different points of time and in relation to particular partnership tasks" [29]: pre-partnership collaboration, partnership creation and consolidation, partnership program delivery and termination/succession.

For the purpose of this paper, the pre-partnership collaboration and the coordination during this phase are of particular interest. Although many renowned and established institutions and networks already work on making restoration a reality, an effective policy network specifically dedicated to the implementation of Aichi Target 15 and overcoming its main challenges has yet to evolve. The pre-partnership phase is "characterized by a network mode of governance based upon informality, trust and a sense of common purpose", which remains essential throughout all phases and is a decisive factor for its success. Empirical analyses have demonstrated the potential of collaborative governance through "goal-directed" networks; on this basis, they are considered a promising governance approach, especially in public sectors, where collective action and "joined forces" are decisive for success [37]. A policy network itself can serve as a starting point for policy implementation. However, while it "may operate through informal patterns of brokerage and shuttle diplomacy", it must eventually develop an explicit and formal strategy to qualify as collaborative governance [27].

3.2. Considerations for Network Creation and Design

Ansell and Gash [27] have identified four factors that influence the potential outcomes of networked governance approaches: starting conditions, institutional design, leadership and the collaborative process. Starting conditions refer to the prehistory of cooperation and conflict that determine the level of existing trust, the power-resource-knowledge relationships between the actors and, eventually, the incentives for, and constraints on, cooperation. Incentives are linked to the actors' expectations and the necessary resources for collaboration: a discernable relationship between individual contributions and tangible outcomes acts as a positive incentive, whereas input limited to advisory or ceremonial purposes is a disincentive [27,38]. In addition, the institutional design and leadership by individual actors have a significant influence on the collaborative process. This process begins with face-to-face dialogue and trust-building that optimally further individual commitments to the process and a shared understanding and eventually result in intermediate outcomes, such as "small wins" and strategic plans for future activities [25].

A network dedicated to achieving a specific target through PPPs can be created through conscious fostering of coordination and cooperation, or it may evolve more spontaneously, when like-minded actors discover the benefits of collaboration for attaining a common goal; in our case, predefined by Target 15. In contrast to markets or hierarchical governance approaches, the governance of the network itself refers mainly to the coordination of its members and actions, since its main feature is its voluntary nature. If a network is actively established, more deliberative decisions can be taken regarding its format. Based on Provan [37], we briefly summarize three different network designs and outline their characteristics, with emphasis on either decentralized self-governance by the network members without a designated governance entity *versus* centralized coordination by a dedicated lead organization elected by the members or through a newly established administrative organization. The suitability of the design depends on different factors, in particular on the purpose, the size, the stage of development, existing relationships and the level of trust between the members [37]. Last, but not least, the degree of consensus regarding the network's purpose and the capability of the network to assemble the required competencies are crucial prerequisites, because they ultimately determine to what extent the individual members will become involved and remain committed to achieving the network's objectives.

The steering of the network and its activities through its members takes place in the form of regular meetings of all members and other, less coordinated efforts. Self-governance requires a high degree of trust and commitment among its members towards the network objective and is marked by symmetrical relationships among its individual entities, which have to manage both internal and external relationships. The advantage of its high flexibility is only exploited if the network remains small in size; the more it grows, the more difficult it is to achieve efficient coordination. With this, the self-governed network faces the choice of either remaining in a "club-like" setting (maintaining its size and excluding new members) or adapting and adopting a new, more centralized form of network governance, where the administration and coordination of member activities are facilitated by a network member or even an external administrative institution. Coordination provided by a network organization results in a more asymmetrical power relationship, which may result in a loss of trust and even lead to the development of rivalries if the organization is not perceived to be neutral or is seen to abuse its function for its own agenda. A solution could be a shared rotating responsibility, as is practiced, for example, within the REDD+ Partnership [39]; this, however, is associated with a notable loss of efficiency, and for the organizations that assume this function, it restricts network activities and opportunities for engagement. It may be an appropriate solution for an evolving network that has grown beyond a size where self-governance is efficient, provided there is an undisputed consensus on an institution that can and wants to assume this administrative role, or it may serve as an interim solution until an external institution is found. Such independent coordination and sustaining of the network and its activities is appropriate when large numbers of participants are involved. Especially when spread over the globe, frequent meetings of all participants become difficult if not impossible; and as a consequence, they either reduce their commitment and participation, which is detrimental for the potential achievement of the network's objectives, or they are required to spend considerable resources on coordination and collaboration. To avoid inefficiency, especially in light of limited resources, it seems that the network governance approach must eventually be brokered, either through a lead organization or through an independent external institution.

Following these theoretical considerations regarding the prerequisites for collaborative governance, we discuss below the challenges to and options for establishing a policy network dedicated to the implementation of Aichi Target 15.

4. A Collaborative Governance Approach for the Implementation of Aichi Target 15

A growing number of studies emphasize that the value of benefits arising from forest restoration exceeds the necessary investments [40], although some assessments of case studies have shown the contrary. Nevertheless, biodiversity protection and restoration activities continue to suffer from chronic underfinancing. The findings have not yet resulted in an adequate mobilization of public funding, which, until today, represents the main funding source for conservation and restoration activities. The endowment of existing bi- and multi-lateral funding sources, such as the Global Environment Facility, can only cover a small fraction of the funding required, especially for activities in developing countries with the largest restoration potentials [1]. Moreover, in times marked by exploding public debts and financial crises, the reiterated call for new and additional public funding remains unheard. This explains why much hope rests on performance-based payments through a REDD+ mechanism currently negotiated under the UNFCCC, e.g., through a "window" for REDD+ in the Green Climate Fund. It could provide funding for large-scale restoration of forests in the context of the eligible activity "enhancement of forest carbon stocks". However, this option is still associated with many uncertainties and is unlikely to materialize before 2020, when the next climate agreement is scheduled to enter into force [41,42]. Another long-demanded option for freeing up existing public funding for restoration is to abolish and redirect subsidies, e.g., for agro-industrial purposes or the use of fossil fuels [13]. Such measures could significantly reduce drivers of land degradation and simultaneously enable large-scale restoration, but the political will for such reforms is lacking.

In recognition of the problems associated with mobilizing new and additional public funding, there is a wide consensus among countries that the private sector must be attracted to and effectively included in the provision of funding (not only for restoration, but also for conservation activities) [13]. Including the private sector, however, creates different, but interlinked challenges. Though seldom explicitly acknowledged, the inherent idea behind this call is that the private sector should become engaged voluntarily, and not through regulatory policy means. There are many motivations for commitment—corporate responsibility, marketing purposes or philanthropy—but commitments must be visible, concrete, simple, efficient and without risk to reputation in order to be attractive to private donors. Another motivation for actors in the private sector is the expectation that a real business case could evolve from investing in forests; given that the potential of philanthropic donors is limited and unlikely to reach necessary levels, it is, on the one hand, desirable to explore such possibilities. On the other hand, creating a business case is associated with considerable risks, as the motivation of most investors is to maximize profits, which has to be balanced with the idea of restoration as a contribution to poverty alleviation (not of the investors, but of locally affected stakeholders). Depending on the degree of degradation, the opportunity costs and other factors, forest restoration can be more costly and is likely to generate less revenue from timber and carbon than investments in commercial tree plantations [43]; investors that prioritize return on investments will try to keep costs as low as possible and maximize revenues.

As addressed in the theoretical framework, collaborative governance through a goal-directed policy network appears to be able to respond to these requirements. In our assessment, we have identified existing networks and initiatives dedicated to promoting forest restoration; their notable achievements so far lie in the brokering of knowledge, the development of tools for practitioners, the promotion of the benefits and the connection of this topic with different political agendas. While this has been successful, the imperative of Aichi Target 15 and the true challenge is to scale up the activities on the ground. For this purpose, it appears inevitable that it will be necessary to tap into a variety of private sources, including investors, which have so far been absent from the networks described. One reason for this is that for many potential investors, it is impossible to distinguish "good" forest projects from those that have been criticized for many reasons, e.g., the "commodification of nature", the disregard for environmental and social aspects, the lack of transparency, and their inherent risks, such as permanence and potential leakage [42,44,45]. In order to dilute existing concerns, a forest restoration business case requires respected and suitable third party certification mechanisms that ensure the environmental and social integrity of the respective activities. Certification is crucial for a number of reasons, but especially in the context of attracting funding, since a major concern of donors and investors alike is protecting their reputation.

In light of these needs, we believe that the existing capacities and initiatives should be bundled through a policy network that functions as a partnership platform and that goes beyond the work of the existing networks: private donors and investors that have so far been largely absent from existing networks need to be attracted and linked into partnerships with those actors that have the knowledge and the capacities to implement forest restoration projects. The different existing networks, partnerships and initiatives (Section 2) demonstrate how many renowned institutions with decades of experience in the field of ecological restoration have effectively organized themselves in different networks, thereby promoting the idea of collaborative governance. These networks serve similar purposes: establishing PPPs, exchanging and brokering knowledge and providing guidance and best practices. Furthermore, they seek to exert influence on decision-makers at all policy levels, and they are very successful in these efforts. Their stated objectives show a wide consensus regarding the benefits of the restoration of degraded landscapes and its contributions to the ecological, economic and social dimensions of sustainable development, for the benefit of present, as well as future generations. In this sense, the expressed missions of the institutions mentioned above reflect a high degree of goal-consensus, a prerequisite for a goal-directed network. This consensus is expressed inter alia in the principles that guide the activities of the institutions described above [26], and also in the degree of trust among the leading actors in these networks, which is indicated through the mutual membership. Moreover, network competencies and know-how, another crucial factor for effective collaborative governance, are available in this case.

Naturally, a policy network has no means to prevent questionable forest investments, but it can and must ensure, through explicit goal-orientated consensus and "social control" through its members, that questionable projects cannot be associated with the network or its objectives. The presence of the strong and well-established networks described above suggests that existing structures can and should be used. A suitable forum for initiating such a policy network would be the follow-up to the Bonn Challenge scheduled for the second half of 2014. Furthermore, the network should include governmental officials of recipient countries that are willing to restore their degraded landscapes and that have the authority to identify priority areas and to help overcome bureaucratic hurdles; in a nutshell, actors who can create an enabling environment for restoration activities. Despite the limited size of the event in 2011, the Bonn Challenge brought together representatives of many key institutions and promoted the idea of private-public partnerships. It has since received much acclaim at high-level policy meetings and has resulted in tangible outcomes and considerable private sector engagement for the implementation of Aichi Target 15 (Section 2). However, with its format as a face-to-face dialogue forum, it can only serve as a starting point for the institutionalization of a policy network that is open to and attractive for new members.

The theoretical considerations suggest some important aspects that should be taken into account when pursuing a collaborative governance approach through such a network. First, in order to ensure effectiveness when a network evolves from a forum and grows, maintaining trust and goal consensus among its actors is a crucial prerequisite. The existing ties and partnerships provide favorable conditions given that all of the institutions described are strongly interlinked and build their work on commonly shared principles that define what ecological landscape restoration actually constitutes. This consensus must be understood and shared by new actors to maintain the level of trust, especially when their core competencies are in fields that extend beyond ecosystem management. Moreover, the idea of creating a business case in addition to philanthropic engagement is found to be worth pursuing. Second, the magnitude of the task creates a requirement for network governance by an external institution, e.g., the GPLFR or the SER; self-governance is inefficient and barely possible. In any case, a small secretariat for facilitating meetings and coordination, as well as the use of modern communication tools should be established. Third, there must be a clear focus on providing incentives for new actors to commit to the network and its activities. This implies that the transaction costs for network participation should be kept at a minimum and allow for tangible contributions to the network's objectives, in concrete implementation projects. For this purpose, and in order to effectively link the network members, the network should establish a partnership platform, which would work like a clearing house mechanism.

A notable example of such a mechanism and its potential is the Life Web platform that was inaugurated at CBD COP9 in Bonn (2008). With its institutional home situated under the roof of the CBD, the Life Web platform was set up to close the immanent funding gap for financing the chronically underfinanced protected areas, particularly, but not exclusively, in developing countries. Its stated mission is "to facilitate financing that helps secure livelihoods and address climate change through supporting the implementation". Recipients (in particular governments) present their funding needs and the relevant information for concrete conservation projects, on a website and in roundtable meetings. Donors, the public and private actors alike can access this information and individually or jointly engage in a highly visible manner in partnerships to implement concrete projects that match their preferences and motivations for engagement. Although focused on the implementation of another (but related) Aichi target with similar funding needs, the "matching platform" of different needs through the Life Web platform could theoretically be broadened and also contribute towards compliance with Aichi Target 15.

5. Conclusions and Outlook

Land suitable for the provision of livelihoods for a rapidly growing global population is limited and already scarce in some regions. Consequently, restoration of degraded lands through the implementation of Aichi Target 15 is an imperative. However, to restore 150 million ha of degraded lands or approximately 20 million ha per year presents extreme challenges, in particular, a considerable need for new and additional funding. Although studies keep emphasizing that the accrued benefits of restoration will outweigh the investments, this policy objective cannot be realized without a joining of public and private forces, given the magnitude of the task and the currently available resources and the persistent problem of tapping into new and additional public funding for conservation and restoration.

In environments marked by resource scarcity and fragmented institutional landscapes, partnerships become "a necessary integrative mechanism" [29] that foster interrelationships, trust and collaboration. Many renowned institutions are already linked in different partnerships and networks related to restoration. They have demonstrated the feasibility of ecological restoration in many projects, have generated a solid knowledge basis and successfully brought the cause to the attention of policy makers at all levels. However, they have not yet sufficiently attracted those actors outside of the conservation community that can make large-scale restoration happen (admittedly a difficult task that requires innovative approaches to secure long-term interest and commitment and that has to be accompanied by high visibility and strong public support). One new and innovative option that will show how far these challenges can be taken up by the private sector may lie in "building forest landscape restoration investment packages". In this context, the first Bonn Challenge in 2011 was very promising. It has resulted in many tangible outcomes and pledges, but the private sector involvement it triggered has so far been insufficient. A repetition of a similar event as planned will provide the chance for public authorities, which depend on the private sector for the implementation of agreed conservation objectives, to initiate a policy network dedicated solely to the shared objective of making Aichi Target 15 a reality. Such a network could help to further streamline the work of the existing institutions in this fragmented, poly-centric institutional landscape [8] and proactively seek to integrate the private sector in financing its implementation; public funding, still the major source of financing, remains insufficient, and despite different options, it appears unreasonable to expect a significant increase in the short term. Attracting private sources and actors with very heterogeneous motivations for such an engagement could be supported by a partnership platform, such as the Life Web initiative, which works like a clearing house mechanism for specific funding needs. Through its high visibility, it creates an incentive for leadership among recipient countries and donors alike; innovative approaches can then be used by successors who can copy-and-paste the format of successful arrangements.

Collaborative governance through a dedicated policy network is a different approach to relying on a hierarchical top-down implementation by public actors alone. While there is no guarantee for its success, existing approaches have not delivered the expected outcomes. With its flexibility, collaborative governance through networks has significant advantages over cumbersome and bureaucratic hierarchies [37]. In light of the global extent of degraded lands, as well as the need to adapt to climate change and to ensure the livelihoods of a growing population, the objective of large-scale landscape forest restoration is a matter of urgency and one that requires innovative approaches in order to be achieved.

Acknowledgments

This study is an outcome of a research project that is financially supported by the German Federal Agency for Nature Conservation, with funds from the German Federal Ministry for the Environment, Nature Conservation and Nuclear Safety (BMUB). The costs associated with this publication were covered by the open access publication fund of the Albert Ludwigs University Freiburg. Special gratitude is dedicated to Emily Kilham for proof reading, to the much appreciated constructive and detailed input provided by the reviewers and last, but not least, to the editors of the special issue.

Conflict of Interest

The authors declare no conflict of interest.

References and Notes

1.   Global Assessment of Opportunities for Restoration of Forests and Landscapes. World Resources Institute, 2011. Available online: http://www.unep-wcmc.org/global-assessment-of-opportunities-for-restoration-of-forests-and-landscapes_765.html (accessed on 28 August 2013).
2.   Geist, H.J.; Lambin, E.F. Proximate Causes and Underlying Driving Forces of Tropical Deforestation. *BioScience* **2002**, *52*, 143–150.
3.   Millennium Ecosystem Assessment. *Ecosystems and Human Well-Being*; Island Press: Washington, DC, USA, 2005.
4.   Leadley, P.; Pereira, H.M.; Alkemade, R.; Fernandez-Manjarrés, J.F.; Proença, V.; Scharlemann, J.P.W.; Walpole, M.J. Biodiversity Scenarios: Projections of 21st Century Change in Biodiversity and Associated Ecosystem Services. In *Technical Series no. 50*; CBD Secretariat: Montreal, Canada, 2010.
5.   Lund, H.G. What Is a Degraded Forest. Available online: http://www.home.comcast. net/~gyde/2009forestdegrade.doc (accessed on 28 August 2013).
6.   Simula, M. Towards Defining Forest Degradation: Comparative Analysis of Existing Definitions. In *FAO Working Paper 154*; FAO: Rome, Italy, 2009.
7.   Braat, L.C.; de Groot, R. The ecosystem services agenda: Bridging the worlds of natural science and economics, conservation and development, and public and private policy. *Ecosyst. Serv.* **2012**, *1*, 4–15.
8.   Restoration is defined as assisting in "the recovery of a degraded, destroyed or damaged forest ecosystem", with the objective "to reestablish a similar pre-disturbance composition of species, structure and functioning conditions of reference ecosystems in the restored area" [46]. Restoration activities always strive to promote synergies between ecological, economic and social aspects [10,13,26], taking into account the specific local circumstances.
9.   Aichi Biodiversity Targets. Available online: http://www.cbd.int/sp/targets/ (accessed on 6 September 2013)
10. Chazdon, R.L. Beyond deforestation: Restoring forests and ecosystem services on degraded lands. *Science* **2008**, *320*, 1458–1460.

11. Sarr, D.A.; Puettmann, K.J. Forest management, restoration, and designer ecosystems: Integrating strategies for a crowded planet. *Ecoscience* **2008**, *15*, 17–26.

12. REDD+ is the UNFCCC acronym for Reducing Emissions from Deforestation and Forest Degradation in developing countries; and the role of conservation, sustainable management of forests and enhancement of forest carbon stocks in developing countries.

13. Lamb, D.; Erskine, P.D.; Parrotta, J.A. Restoration of degraded tropical forest landscapes. *Science* **2005**, *310*, 1628–1632.

14. Easterly, W. Institutions: Top down or bottom up? *Am. Econ. Rev.* **2008**, *98*, 95–99.

15. FAO; JRC. Global Forest Land-use Change 1990–2005. FAO Forestry Paper No. 169, 2012. Available online: http://www.fao.org/forestry/fra/fra2010/en/ (accessed on 9 September 2013).

16. Butchart, S.H.M.; Walpole, M.; Collen, B.; van Strien, A.; Scharlemann, J.P.W.; Almond, R.E.A.; Baillie, J.E.; Bomhard, B.; Brown, C.; Bruno, J.; *et al.* Global biodiversity: Indicators of recent declines. *Science* **2010**, *328*, 1164–1168.

17. Mace, G.M.; Cramer, W.; Díaz, S.; Faith, D.P.; Larigauderie, A.; Le Prestre, P.; Palmer, M.; Perrings, C.; Scholes, R.J.; Walpole, M.; *et al.* Biodiversity targets after 2010. *Curr. Opin. Environ. Sustain.* **2010**, *2*, 3–8.

18. Andresen, S.; Hey, E. The Effectiveness and legitimacy of international environmental institutions. *Int. Environ. Agreem. Polit. Law Econ.* **2005**, *5*, 211–226.

19. Talberth, J.; Grey, E. Global Costs of Achieving the Aichi Biodiversity Targets—A Scoping Assessment of Anticipated Costs of Achieving Targets 5, 8, and 14. 2012. Available online: http://www.cbd.int (assessed on 9 October 2013).

20. UNEP/CBD/COP/11/14/Add.2: Report of the high-level panel on global assessment of resources for implementing the strategic plan for biodiversity 2011–2020; 2012. Available online: http://www.cbd.int (accessed on 9 October 2013).

21. Hein, L.; Miller, D.C.; de Groot, Rudolf. Payments for ecosystem services and the financing of global biodiversity conservation. *Terr. Syst.* **2013**, *5*, 87–93.

22. Bonn Challenge and Landscape Restoration. Available online: http://www.iucn.org/about/work/programmes/forest/fp_our_work/fp_our_work_thematic/fp_our_work_flr/more_on_flr/bonn_challenge/ (accessed on 9 October 2013).

23. Landscapes for A Sustainable World. Available online: http://www.landscapes.org/ (accessed on 9 October 2013).

24. Giessen, L. Reviewing the main characteristics of the international forest regime complex and partial explanations for its fragmentation. *Int. For. Rev.* **2013**, *15*, 60–70.

25. Melo, F.P.L.; Pinto, S.R.R.; Brancalion, P.H.S.; Castro, P.S.; Rodrigues, R.R.; Aronson, J.; Tabarelli, M. Priority setting for scaling-up tropical forest restoration projects: Early lessons from the Atlantic Forest Restoration Pact. *Environ. Sci. Policy* **2013**, *33*, 395–404.

26. Sayer, J.; Chokkalingam, U.; Poulsen, J. The restoration of forest biodiversity and ecological values. *For. Ecol. Manag.* **2004**, *201*, 3–11.

27. Ansell, C.; Gash, A. Collaborative governance in theory and practice. *J. Public Adm. Res. Theory* **2008**, *18*, 543–571.

28. Imperial, M. Using collaboration as a governance strategy: Lessons from six watershed management programs. *Adm. Soc.* **2005**, *37*, 281–320.
29. Lowndes, V.; Skelcher, C. The dynamics of multi-organizational partnerships: An analysis of changing modes of governance. *Public Adm.* **1998**, *76*, 313–333.
30. Mert, A. Partnerships for sustainable development as discursive practice: Shifts in discourses of environment and democracy. *Discourse Expert. For. Environ. Gov.* **2009**, *11*, 326–339.
31. Hale, T.N.; Mauzerall, D.L. Thinking globally and acting locally: Can the Johannesburg partnerships coordinate action on sustainable development? *J. Environ. Dev.* **2004**, *13*, 220–239.
32. Young, O.R. Regime dynamics: The rise and fall of international regimes. *Int. Organ.* **1982**, *63*, 277–297.
33. Teisman, G.R.; Klijn, E.H. Partnership arrangements: Governmental rhetoric or governance scheme? *Public Adm. Rev.* **2002**, *62*, 197–205.
34. Visseren-Hamakers, I.J.; Arts, B.; Glasbergen, P. Interaction management by partnerships: The case of biodiversity and climate change. *Glob. Environ. Polit.* **2011**, *11*, 89–107.
35. Pattberg, P.; Biermann, F.; Chan, S.; Mert, A. Introduction: Partnerships for Sustainable Development. In *Public Private Partnerships for Sustainable Development: Emergence, Influence and Legitimacy*; Pattberg, P., Biermann, F., Chan, S., Mert, A., Eds.; Edward Elgar Publishing: Cheltenham, UK, 2012; pp. 1–18.
36. Pattberg, P. The Role and Relevance of Networked Climate Governance. In *Global Climate Governance beyond 2012: Architecture, Agency and Adaptation*; Biermann, F., Pattberg, P., Zelli, F., Eds.; Cambridge University Press: Cambridge, UK; pp. 146–164.
37. Provan, K.G.; Kenis, P. Modes of Network Governance: Structure, Management, and Effectiveness. *J. Public Adm. Res. Theory* **2008**, *18*, 229–252.
38. Futrell, R. Technical adversarialism and participatory collaboration in the US chemical weapons disposal program. *Sci. Technol. Hum. Values* **2003**, *28*, 451–482.
39. Reinecke, S.; Pistorius, T.; Pregernig, M. UNFCCC and the REDD+ Partnership from a networked governance perspective. *Environ. Sci. Policy* **2014**, *35*, 30–39.
40. Bullock, J.M.; Aronson, J.; Newton, A.C.; Pywell, R.F.; Rey-Benayas, J.M. Restoration of ecosystem services and biodiversity: Conflicts and opportunities. *Trends Ecol. Evol.* **2011**, *26*, 541–549.
41. Pistorius, T. From RED to REDD+: The evolution of a forest-based mitigation approach for developing countries. *Curr. Opin. Environ. Sustain.* **2012**, *4*, 638–645.
42. Gutman, P.; Davidson, S. A Review of Innovative International Financing Mechanisms for Biodiversity Conservation with a Special Focus on the International Financing of Developing Countries' Protected Areas. WWF International: Washington, DC, USA, 2007. Available online: http://www.conservation.org (accessed on 4 March 2011).
43. Olsen, K.H. The clean development mechanism's contribution to sustainable development: A review of the literature. *Clim. Chang.* **2007**, *84*, 59–73.

44. Schlamadinger, B.; Bird, N.; Johns, T.; Brown, S.; Canadell, J.; Ciccarese, L.; Dutschke, M.; Fiedler, J.; Fischlin, A.; Fearnside, P.; *et al.* A synopsis of land use, land-use change and forestry (LULUCF) under the Kyoto Protocol and Marrakech Accords. Options for including agriculture and forestry activities in a post-2012 international climate agreement. *Environ. Sci. Policy* **2007**, *10*, 271–282.

45. Kickert, W.; Klijn, E.-H.; Koppenjan, J. Introduction: A management perspective on policy networks. In *Managing Complex Networks: Strategies for the Public Sector*; Kickert, W., Klijn, E.-H., Koppenjan, J., Eds.; Sage: London, UK, 1997.

46. Society for Ecological Restoration. The SER Primer on Ecological Restoration, 2004. Available online: http://www.ser.org (accessed on 6 September 2013).

Multi-Scalar Governance for Restoring the Brazilian Atlantic Forest: A Case Study on Small Landholdings in Protected Areas of Sustainable Development

Alaine A. Ball, Alice Gouzerh and Pedro H. S. Brancalion

Abstract: Implementation of forest restoration projects requires cross-scale and hybrid forms of governance involving the state, the market, civil society, individuals, communities, and other actors. Using a case study from the Atlantic Forest Hotspot, we examine the governance of a large-scale forest restoration project implemented by an international non-governmental organization (NGO) on family farmer landholdings located within protected areas of sustainable development. In addition to forest restoration, the project aims to provide an economic benefit to participating farmers by including native species with market potential (fruits, timber) in restoration models and by contracting farmers in the planting phase. We employed qualitative methods such as structured interviews and participant observation to assess the effect of environmental policy and multi-scalar governance on implementation and acceptability of the project by farmers. We demonstrate that NGO and farmer expectations for the project were initially misaligned, hampering farmer participation. Furthermore, current policy complicated implementation and still poses barriers to project success, and projects must remain adaptable to changing legal landscapes. We recommend increased incorporation of social science methods in earlier stages of projects, as well as throughout the course of implementation, in order to better assess the needs and perspectives of participants, as well as to minimize trade-offs.

Reprinted from *Forests*. Cite as: Ball, A.A.; Gouzerh, A.; Brancalion, P.H.S. Multi-Scalar Governance for Restoring the Brazilian Atlantic Forest: A Case Study on Small Landholdings in Protected Areas of Sustainable Development. *Forests* **2014**, *5*, 599-619.

1. Introduction

Tropical forest regions, as sites of both high biodiversity and high rates of ecosystem transformation and degradation, are a focus of conservation and forest restoration initiatives worldwide [1]. To effectively achieve multiple objectives of biodiversity conservation, forest restoration, and sustainable development throughout these regions, multi-scalar governance systems that engage state and non-state actors across levels of governance are required [2]. As large-scale environmental issues faced by all mankind, such as climate change and biodiversity loss, increase in complexity, so too do our understandings of the range of solutions and partnerships necessary to address these problems. With this understanding comes a recognition that no one sphere offers the best approach, but that strategies require the cooperation, interaction, and interdependence of different sectors. When effective, these interdependencies comprise systems of "good governance".

In recent decades, "good governance" has gained popularity in environment and development as a mechanism with which to improve management of economic, social, and environmental resources [3–5]. Like other concepts employed in development, such as "participation" and "community-based natural resource management (CBNRM)," "good governance" has become 'institutionalized' and normative in theory and practice of socially just development, despite broad interpretation of its meaning [4,5]. Indeed, good governance is considered essential for fair and multi-level resource management and should employ the following principles: openness, participation, accountability, effectiveness, coherence, and civic peace [6,7]. As with other development concepts, practice of good governance is closely linked with ideals of democracy and with market mechanisms (both private and state-driven) for addressing rural poverty [8–10], but as a model remains necessarily undefined in order to be applicable to diverse local and institutional contexts [4].

An emerging and promising field of governance studies and theory, environmental governance is a concept encompassing all forms of action, organization, and formal and informal rule-making directed at addressing matters of the environment, especially environmental problems. Lemos and Agrawal [11] describe "environmental governance" as "the set of regulatory processes, mechanisms and organizations through which political actors influence environmental actions and outcomes" and place emphasis on the effectiveness of "hybrid" *versus* "pure" modes of governance, such as state-only or market-only solutions. Instead, cross-scale and co-governance partnerships between state, market, civil society, individual, community, and other actors offer increased opportunity for information exchange, adaptive management, and access to knowledge, benefits and authority [11,12]. Adaptive governance, as described by Folke *et al.* [13], allows the partnerships and management systems crucial to environmental governance to respond to changing social, economic, and ecological conditions, enhancing the resilience [14] of systems being governed.

Civil society [15,16] can play crucial roles in addressing environmental problems and improving democratic participation, enhancing good governance of resources. Non-governmental organizations (NGOs), labor unions, and local associations and cooperatives may improve smallholder access to benefits such as credit, technology, information, and markets, advancing their ability to participate in governance of production systems and, in this case, of forest restoration. Indeed, NGOs are often able to more directly address the needs of rural populations due to their greater flexibility or by acting as intermediaries between households, governments, funders, and the private sector [17]. However, the presence of such organizations is not a guarantee of success, and a correlation between NGO intervention and expanded "political spaces" for the poor cannot be assumed [18]. Like all institutions [19,20], those of civil society are subject to the effects of conflicting interests and management challenges, particularly relevant in the multi- and trans-disciplinary field of forest restoration. Furthermore, civil society organizations often work at the "community" level, resulting in problematic homogenization of diverse local conditions [21,22], and seek "win-win" solutions rather than addressing realistic trade-offs [23].

As defined today, forest restoration engages ecological and social systems to modify landscapes, ecosystem processes, and people, dependent upon the interests, interactions, and capabilities of multiple actors [24]. Actors can include forest restoration specialists with training as ecologists, biologists, foresters, and technicians; federal, state, and local government agencies; financial institutions (both public and private); civil society (NGOs, local associations and cooperatives); private businesses and industrial sectors seeking to establish themselves as progressive and "green"; and rural communities. Because all actors operate across multiple spheres of authority and knowledge, forest restoration projects necessarily involve cross-scale formal and informal arrangements of governance, as well as systems to be governed [25]. Transcendence of territorially bounded conceptions of governance permits engagement with "new political spaces" in a non-hierarchical manner, with important implications for political asymmetry and power sharing among actors [2]. As mentioned, civil society can play crucial roles in negotiating this asymmetry.

As opposed to purely conservationist approaches, forest restoration has since its early stages recognized the importance of social systems in the forest restoration process [26]. However, early conservationist perspectives that considered local populations as destroyers of the environment and excluded them from management strategies have bred conflicts that continue to pose challenges to implementation of forest restoration projects [27]. As a result, conservation, and sometimes forest restoration, projects have often resulted in failures and been considered as "neocolonialist" [27,28]. More recently, socially minded ecological restoration has been described as restoration of natural capital (RNC). RNC is a concept that considers the interface between ecology and economics, and between people and the natural environment, drawing on various disciplines including social sciences, economics, and policy. It suggests the necessity to develop a more holistic approach and accentuate the consideration of historical, political, economic and cultural factors for forest restoration projects to succeed [29–32]. Forest restoration is now a truly multidisciplinary field of action.

RNC was built on the idea that forest restoration should operate beyond purely technical and scientific knowledge and engage people in the forest restoration process, and that compromised natural capital is a limiting factor for human well-being and economic sustainability [33]. Traditional populations, family farmers and small landowners have an invaluable experiential knowledge about their environment and often contribute to the sustainable management of natural resources, thus it is increasingly suggested that they should take part in the design of conservation and forest restoration projects [34]. The hypothesis that traditional populations may contribute to conservation effectiveness was considered in the work of Porter-Bolland *et al.* [35], which suggested that community managed forests distributed across the tropics showed lower deforestation rates than strictly protected areas. Such studies imply that when engaging traditional and family farmers, forest restoration practitioners could incorporate local environmental knowledge and local management techniques into project design, potentially increasing project success.

Due to the cross-disciplinary nature of forest restoration, careful observations about the operation of governance of forest restoration projects are necessary to improve the design, implementation, and success of forest restoration. Using a case study from São Paulo State, southeastern Brazil, we address institutional project management by public, private, and civil society bodies; public policy; and multi-scalar implementation in a large-scale forest restoration initiative. The studied project is being currently implemented by an international NGO on small landholdings located in protected areas of sustainable development of the Atlantic Forest Hotspot [36].

Working from a framework of trade-offs rather than win-wins, we set reasonable expectations for project successes as well as gain a realistic picture of social, ecological, and political realities. To map these realities and assess the governance systems of our case study, we ask,

- How did the governance of this forest restoration case study by a large NGO and the current legal context affect the project's implementation and acceptability by farmers?
- What are the relationships between local "community" and institutional-level governance, and how do they affect project success?

Previous studies have described the multi-scalar nature of sustainable development [10,25], both promoted and critiqued the concepts of CBNRM and co-management between the state and communities [37], and addressed development trade-offs [23]. However, the relationships among actors, across scales, and between policy and implementation of forest restoration projects remain understudied. Furthermore, mechanisms of actor relationships in the context of social-ecological relationships are not well understood, problematizing recommendations for increased resilience in systems of environmental governance [12]. We seek to provide a clearer picture of these relationships by examining a forest restoration project as an "object" of governance, with the intention of offering insight into improved implementation of forest restoration initiatives involving smallholders.

2. Experimental Section

2.1. Study Sites and Project History

This study was carried out in the Atlantic Forest of Brazil, a global biodiversity Hotspot and, more specifically, in the Serra do Mar biogeographical sub-region, the best-preserved center of endemism of this biome [38]. For achieving the goals of this research, we chose as a case study a forest restoration project implemented in the municipality of Barra do Turvo, Vale do Ribeira region (Figure 1; Detailed ecological, socioeconomic and land use information can be found in Table 1). Funded by the Brazilian Development Bank (BNDES, *Banco Nacional do Desenvolvimento*) as part of its Atlantic Forest Initiative, the project also seeks to provide an economic incentive for farmers to participate in the forest restoration process. The Atlantic Forest Initiative allocates funding for the implementation of forest restoration projects across the biome, and the NGO responsible for the implementation of the studied project received approval from the Bank to include the economic component in the project design. Forest restoration is being conducted on farmer property located within Sustainable Development Reserves (RDS, *Reserva de Desenvolvimento Sustentável*), a category of protected area that permits management, and native species with potential for farmers to exploit economically via fruits and timber are favored.

Most farmers of Barra do Turvo are from traditional groups such as the *Caiçaras* and the *Quilombolas* [40] or are considered "family farmers" and live from a combination of subsistence agriculture, banana production, and the extraction of natural resources of the forest, such as the emblematic *palmito juçara* (*Euterpe edulis*) and its "heart of palm," which is threatened with extinction due to overexploitation [41]. It is for these reasons and others outlined in Table 1 that the NGO targeted this region for a project designed to bring environmental, economic and social benefits.

Figure 1. Localization of the Mosaic of Jacupiranga (MOJAC) and of the study sites in Barra do Turvo, southeastern Brazil, where governance issues were assessed for a forest restoration program carried out on small landholdings in Protected Areas of Sustainable Development. Modified from "Map of the Mosaic of Jacupiranga," Instituto Socioambiental, 2008; [39].

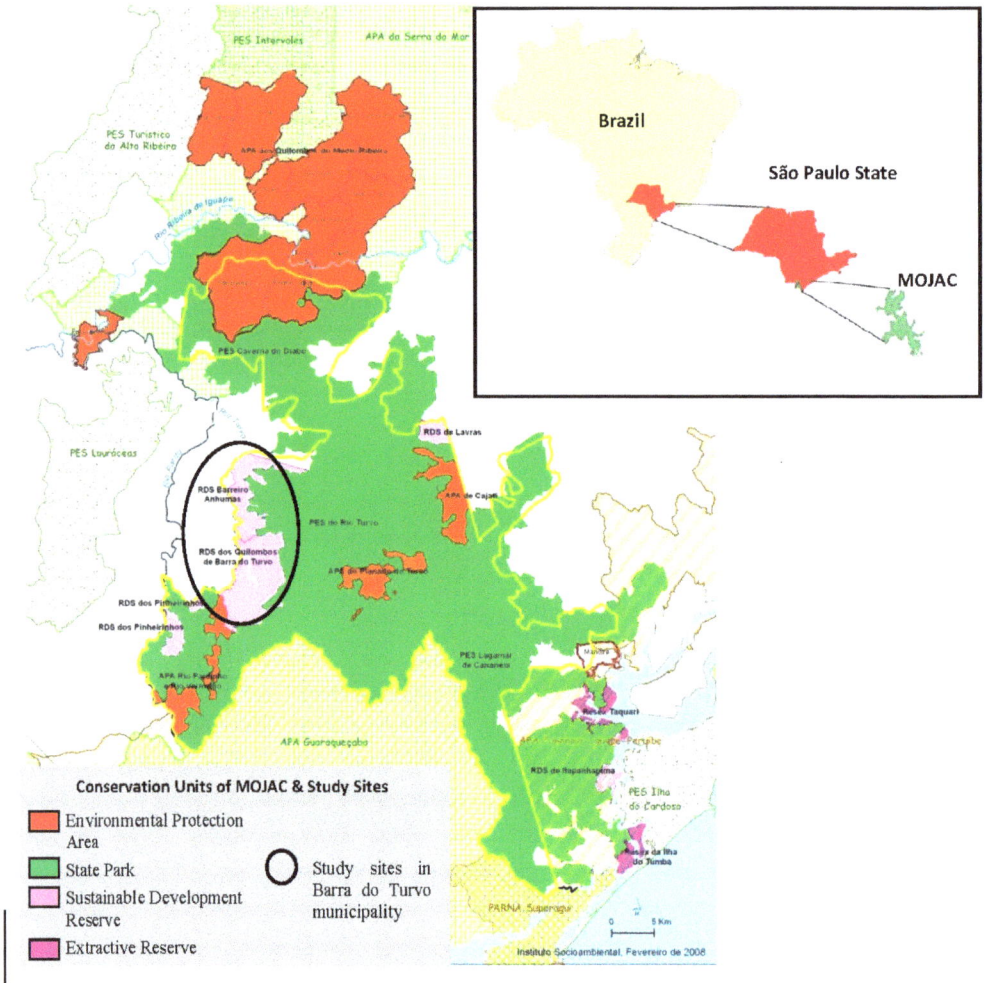

Table 1. Ecological and socioeconomic descriptions of the study sites, considering different scales, where governance issues were assessed for a forest restoration program carried out on small landholdings in Protected Areas of Sustainable Development in Barra do Turvo, southeastern Brazil.

	Biome-scale	Regional-scale	State-scale	Local-scale
The study site	Atlantic Forest	Serra do Mar	Vale do Ribeira	Barra do Turvo
Ecological	A global biodiversity Hotspot that once covered 1.5 million ha, but is now reduced to 12% of its original cover. Human-modified landscapes predominate.	This biogeographical sub-region is one of the centers of endemism of the Atlantic Forest, retaining 36.5% of its original vegetation. It contains the three largest remnants of continuous forest of the biome and accounts for 63% of the total remaining Atlantic Forest under protection.	Contains 2 million ha of forested areas (21% of the total Atlantic Forest), which shelter great biological richness and potential for sustainable management, such as through agroforestry or ecotourism. One of the most threatened plant species of this region, and also one of the most economically exploited for palm heart, is *Euterpe edulis*.	48% of the municipality's total surface is covered by some of the largest remnants of native forest (dense ombrophylus rainforest). But, in recent decades, farmers have witnessed soil fertility and water quality decrease, and degradation and erosion escalate, due to intensified grazing and use of fire as fewer areas are available to exploit.
Socioeconomic	This biome harbors more than 60% of Brazil's population within its boundaries, where *ca.* 70% of the national GDP is generated.	Several prosperous, big cities are located in this region, with many industries and services. However, small, poor cities predominate in the most forested regions, where agriculture is the main source of income of the population.	The region is remote and poorly linked to the state's main cities. With the lowest Human Development Index in the state and a very low population density, it is the poorest region of São Paulo State.	The commercialization of small farmers' products is still uncertain, and the intensification of agriculture limited by the hilly relief. Farmers mainly rely on bananas and peach palm (*Bactris gasipaes*) to generate income, as well as on cattle, which acts as a "security capital" used in times of necessity. Their situation has improved with the creation of cooperatives and the establishment of governmental "Food Acquisition Programs".
Land use	Mostly urbanization, extensive pastures and intensive agriculture (sugarcane, eucalyptus, orange).	Principal land uses include reserves (protected areas cover 25.2% of the region) and extensive pasturelands with very low productivity. Remaining forests are highly explored by the population.	Intensive and small-scale banana production as well as subsistence agriculture. Cattle ranching is also a major land use, particularly on steep slopes. Forest management is also very common.	The production systems are varied: Agroforestry, native fruits, cattle and buffalo breeding, crops, vegetables and beans. Pastures are a major component of the landscape. With the exception of bananas, peach palm, and milk, production is mainly for household consumption.

In search of alternatives, conscious of the forest's values, and with the need to adapt to environmental laws and rules of the Mosaic of Jacupiranga (*Mosaico do Jacupiranga*, MOJAC) [42], farmers have incorporated forest restoration and agroforestry as new activities (Table 1). Overall, the farmers of the municipality are eager to experiment with new production systems, thus they initially welcomed the NGO. There are two RDS in Barra do Turvo: The RDS Quilombos de Barra do Turvo, constituted by four Quilombos, or traditional communities, with a total of 136 families; and the RDS Barreiro-Anhemas, constituted by two neighborhoods with a total of 176 families of family farmers. Considering that the NGO designed the project to be conducted on 21 hectares in this municipality, a total of sixteen farmers from various communities of both RDS joined the project.

To realize the project within the RDS, the NGO partnered closely with the Forest Foundation (FF, *Fundação Florestal*), a government body of the Secretariat of the Environment of São Paulo that is responsible for the management of state protected areas. Local FF RDS managers, along with an NGO technician later hired for local project management, were responsible for the presentation and coordination of the project with communities. However, allocation of funds was delayed until 2012, two years after initial discussion with farmers and the FF, and many farmers lost faith in or forgot about the project during this time. After implementation was reinitiated in early 2013, the NGO contracted a forest restoration consulting company and the biggest nursery of native species in the state, and the University of São Paulo's (USP) Laboratory of Ecology and Forest Restoration (LERF, *Laboratório de Ecologia e Restauração Florestal*) and Laboratory of Tropical Forestry (LASTROP, *Laboratório de Silvicultura Tropical*) to design forest restoration models in conjunction with communities for the chosen areas.

After initial design of forest restoration models at USP, a participatory workshop was held at an RDS headquarters at which the NGO, consulting company, and LERF/LASTROP presented to farmers a model of "sequential planting," in which pioneer species are planted first, followed by later successional species in subsequent years. Native species to be planted were determined jointly with farmers, and *E. edulis* was agreed upon as the species with greatest future economic benefit through the use of its fruits to produce a pulp similar to that of *açaí*, *Euterpe oleracea*. Additionally, a daily rate will be paid to farmers who assist in the planting phase, though the NGO was not able to provide the amount of this rate at the time of the workshop. The NGO will provide technical assistance to farmers for maintenance of the forest restoration sites for a period of two years.

2.2. Methods

The study was realized at the time of the implementation phase of the project, between May and August 2013. The researchers were part of the field team collaborating with the NGO, in charge of prospecting 21 hectares to be restored in Barra do Turvo, and used this opportunity to familiarize themselves with the study sites and the different actors involved in the project. Five fieldtrips (approximately twelve days total) were organized to Barra do Turvo to meet the RDS manager and interested farmers, explain the project and realize the environmental diagnosis of the areas. These trips also enabled the researchers to conduct short preliminary interviews with the farmers. The

NGO's technician in charge of the project in São Paulo accompanied the team in the field each time it was possible. Additional data was collected through participant observation during a one-day workshop organized in August gathering all stakeholders and through review of project documents provided by the NGO. We also investigated the legal instruments and regulations affecting the realization of the project since its beginning.

Eighteen separate structured interviews with farmers of fourteen distinct households (from both RDS) were conducted exclusively for the study without the presence of the NGO during a seven-day fieldtrip in July 2013. Thirteen of the interviewees are participating in the project and the five others had declined the NGO's offer. The objective of the interviews was first to realize a brief agrarian system diagnosis of the region and of each household, which improved understanding of the farmers' practices, their involvement in the community, the difficulties encountered, the role of each production system and the cash flow. This step, which had not been realized by the NGO, was essential to better appreciate farmers' enthusiasm or reluctance towards the project. Indeed, the history of interventions conducted for agricultural development reveals that the actions taken cannot be effective without knowing beforehand the dynamics of the agrarian system and the diversity of production systems of the region [43]. Interview questions addressed tenure, daily on-farm activities and major crops, labor and materials available, changes in focal activities over time, and other income sources apart from farming.

Then we investigated the evolution of the relationship between the farmers and the forest over time, the activities linked to it and the idea and opinion the farmers have of reforestation. Interview questions focused on the role the forest and trees play on farmer property and in production systems, observed environmental changes over time, and understanding of forest restoration and of the current project. Engagement with local farmer associations and cooperatives was also assessed, as were opinions and perceptions of the current project, including how and why the farmer became involved. All this information gave insight into the values the farmers associate with this ecosystem and their expectations about the project.

Finally, additional interviews with the NGO's former Project Manager and current Program Manager completed our effort to better understand the project's history, the governance system and the barriers encountered, as well as the NGO's own vision and expectations about the farmers and the project. In these interviews, we asked about the NGO's relationship with BNDES and how the project was revised to include an economic benefit to farmers, choice of areas to be restored and relationship with the FF, how the NGO perceives its relationship with the farmers, and how it perceives farmer understanding of the project, and difficulties in implementation.

The analysis of the data collected during field surveys, interviews and participant observation is entirely qualitative. The information was coded and sorted according to our research questions into thematic groups.

3. Results and Discussion

3.1. Institutional Project Management

3.1.1. Policy Context and Constraints

The legal and tenure conditions in which the project takes place are complex, placing constraints on project implementation and success at later stages. Major legislation affecting the project include the recently revised and heavily debated Brazilian Forest Code, the Atlantic Forest Law, the National System of Units of Conservation (SNUC, *Sistema Nacional de Unidades de Conservação*), and legislation regulating the management and commercialization of native, endangered species (Table 2).

In all areas of the project, sites are carved into micro-divisions, each with associated legal and prohibited activities. For conservation, the Forest Code defines two types of areas: The Areas of Permanent Preservation (APP, *Áreas de Preservação Permanente*) and the Legal Reserve (RL, *Reserva Legal*). The APPs are riparian areas and steep slopes that cannot be exploited for economic activities, and the size of an APP varies according to the width of the river and the size of the landholding considered. Law mandates compulsory forest restoration of APPs when they are in a degraded state, however, with the new Forest Code of 2012, the size of APPs to be restored has been greatly decreased, with just the region closest to water bodies now obligatory to recuperate (the rest is called "consolidated" and can be managed by the property holder). Changes in the new Code underscore a key difficulty of planning such a project: The necessity to design the best schema in accordance with current legislation while anticipating future changes in legislation that will directly affect how the project will operate. It must work within the current legal framework while hoping for changes conducive to success, such as policy favorable to management of secondary forest and of *E. edulis* fruits. Because of the uncertainty of this scenario, projects and local managers must maintain flexibility in implementation over time, especially when the forest restoration project is focused on the exploitation of long-lived native species.

Several other policies affect the management regimes of the project. The SNUC is a governmental instrument created in 2000 to work towards the protection of the environment and guarantee the right of traditional populations to access resources necessary for their subsistence. SNUC defines two groups of Units of Conservation with specific characteristics and objectives, as presented in Table 2. The primary objective of an RDS (group of Sustainable Use) is sustainable management of the reserve in order to preserve both biodiversity and the local communities' knowledge and traditions, as well as to increase their quality of life. Like all of São Paulo State's protected areas, the RDS is a public domain administrated by the Forest Foundation. By allowing sustainable management in RDS, SNUC makes the economic component of the studied project possible.

Table 2. Legal instruments and implications for the development of a forest restoration program carried out on small landholdings in Protected Areas of Sustainable Development in Barra do Turvo, southeastern Brazil.

NAME		TARGET	OBJECTIVES	IMPLICATIONS FOR PROJECT DEVELOPMENT
Brazilian Forest Code	Federal Law 12.651/2012	Forest protection and restoration	→ Regulates land-use: - APP: protected areas (riparian areas, slopes >45°, hilltops, any areas >1800 m), divided in obligatory (must remain vegetated) and consolidated (ability to manage). - RL: sustainable exploitation allowed but requirement to maintain or restore the forest (80% of the property in Amazon Forest; 35% in Savannas; 20% in Atlantic Forest).	→ Defining areas to be restored. → Management dependent upon government authorization and licensing. → Right to manage agroforestry systems in RL and consolidated APP by "traditional" or "small family farmer". Raises the issue of being recognized as such; imposes limits to farmers and to the technical nature of the project (practices, species or resources targeted).
Atlantic Forest Law	Federal Law 11.428/2006	Atlantic Forest remnants	→ Regulates vegetation removal based on the stage of regeneration (strongly restricted in secondary forests)	→ Defining areas to be restored → Management dependent upon government authorization and traditional status.
National System of Units of Conservation (SNUC)	Federal Law 9.985/2000	Environment and traditional populations' rights	→ Protects the environment and guarantees access to lands and natural resources for traditional peoples: - Group of Full Protection: Conservation, research, education -Group of Sustainable Use: Conservation, sustainable management.	→ Defining traditional status and management activities → Land tenure regulated by the Conservation Unit's management plan, raising the issue of written title versus customary tenure. → Regulating exploitation of native species.
MMA/IBAMA	Federal Law 6/2008	Species threatened with extinction	→ Protection and improvement of native flora and fauna (restrictions on extraction, conservation ex-situ).	→ Regulating exploitation of native species.
SMA	State Law48/2004	Species threatened with extinction in São Paulo State	→ Protection and improvement of native flora and fauna	→ Regulating exploitation of native species.
National System of Seeds and Seedlings	Federal Law 10.711/2003	Seeds and seedlings	→ Organizes production, trade, import and export system; assurance of identity and quality	→ Regulating production of native species

Finally, another difficulty for a forest restoration project such as this one—as it also aims to provide economic benefits—is to obtain the right to work with protected or threatened species. *Euterpe edulis* is an endangered species increasingly favored in projects, as it is crucial in the ecology of the forest and can provide income for farmers via seeds or processing of fruit pulp [41]. Other uses such as the extraction of its heart of palm are forbidden, even if the palm tree was originally planted by a farmer, as the process leads to the death of the palm [44]. This problem is encountered for numerous native species, and a whole project faces failure if it cannot guarantee the farmers authorizations to manage and use the resources, and consequently reduces their willingness to plant native tree species in agricultural lands.

Farmer uncertainty regarding future ability to benefit economically from native species was not only a function of difficulty on the part of the NGO in explaining the economic component of the project, but also a reflection of the reality of policy complexity surrounding native species. Farmers are fully aware of restrictions on commercializing, or even cutting for personal use, native timbers, and future changes in these restrictions are uncertain. In an encouraging development, São Paulo State recently adopted a resolution regulating management of *E. edulis* fruits (SMA 105/2013), but while most participants in the project are hopeful about pulp commercialization, they realize that this represents a long-term benefit from which they cannot immediately profit (the *E. edulis* palm typically begins producing fruit only eight years after planting). If legislation does not facilitate other native species management in the future, or even the cultivation of crops in the initial phases of the project through agro-successional models of forest restoration [45], these restrictions could in fact prevent farmer access to manage trees they have planted for this project.

Presently, the law permits developing agroforestry or agro-successional systems to be managed by traditional or small family farmers in the consolidated APP and RL. Management may also be permitted in young secondary forests provided it is for subsistence use. But undoubtedly, the project is embedded in a complex legal landscape where it is hard to know which law prevails, and how future legislation will support or hinder project objectives.

Management rights are also complicated by the location of farmer properties within protected areas of sustainable use. As government property, RDS land is subject to regulations defined in the reserves' management plans and by the FF. Furthermore, before transfer of the land to the government, the majority of farmers held only *posse* (possession through long-term inhabitance) rather than written title, further weakening their property claims. Especially for the Quilombo communities concerned in this study, title remains a point of contention between communities and the FF. All farmers' right to remain living and producing within the RDS is contingent upon their identity as "small" or "traditional," defined by size of property and on-farm methods.

3.1.2. Incentives and Project Acceptability

The prospect of future economic gains from planted species may have offered additional incentive for farmers to participate [46] but was not found to be the principal reason for acceptance of the project. Rather, farmers were more likely to participate if they simply had marginal lands not currently in use and perceived no detrimental effect of allowing forest restoration on their property. Farmers with cattle or buffalo, whose forest restoration areas will require construction of fences, were enthusiastic about receiving new fencing through the project, and farmers whom the project will employ for area preparation, planting, and maintenance agreed to participate because of the income provided by these activities. Yet, even when farmers displayed interest in experimenting with a new species or management technique, they were not always willing or able to invest time or money in this experimentation.

Furthermore, interviews indicated that farmers do not currently hold a vision of the forest as a source of economic benefit, as Brazilian environmental law has largely rendered it off-limits to management. While some farmers rely on forest products for traditional use, income is primarily dependent on non-forest production systems. Thus, the majority of interviewees do not believe that any economic benefit from the project will significantly increase their income. Prior experience with or exposure to forest restoration had not involved any potential for smallholders involved in this study to economically exploit species planted in the future, so conceptually this was a very difficult idea to convey when explaining the project to farmers. Interviews made evident the fact that, although the NGO had previously introduced the concept of the project to them during the first visits in 2010 and 2011, nearly all respondents were totally unclear about the idea that the forest restoration model adopted was meant to provide a future economic benefit for them.

From the start, NGO and farmer understandings of the project were not in alignment. Before and during the area diagnostic and mapping phase, the project faced many setbacks as participants dropped out, unsure of the intended benefits of the project to them and distrustful of the NGO's intentions after a long delay in implementation with no communication with participants during the period of the delay. Property visits by field staff and the workshops held by the NGO and the FF greatly contributed to farmer understanding of the NGO's vision, and participants were enthusiastic about the future potential for *E. edulis* pulp. Before the workshops, half of our informants described reforestation as "planting trees on an area you can't use anymore afterward" and as something that is "using up space" and a "loss of agricultural lands." These statements underline the smallholder perception of reforestation on their land as a loss of usable space, either for cattle or crops, and of a use, rather than conservationist, relationship with the landscape.

Despite this use-based relationship, during workshops and interviews, farmers cited many ecosystem service values, such as the provisioning and regulating services of recovery and maintenance of soil fertility, fresh water, and air quality, the cultural service of inherent beauty, and the supporting service of animal habitat [TEEB service categories; 1]. Articulation of ecosystem service values of the forest by farmers demonstrates a shared value with NGOs, funders, and environmental policy and serves as a point of mutual understanding of the benefits of a forest restoration project. By becoming more familiar with the association between ecosystem services and forest restoration, and by witnessing increased economic potential for native species, farmer goals will become progressively more aligned with those of forest restoration [33]. Furthermore, projects should place greater emphasis from the start on arriving at mutually understandable definitions of key concepts, such as the definition of forest restoration itself, in order to ensure successful implementation and avoid later confusion between stakeholders [47].

3.2. Multi-Scalar Implementation

3.2.1. Participatory Nature of the Project

The implementation of a large-scale forest restoration project funded by BNDES, designed by a multi-national and hierarchical NGO, and ostensibly intended to benefit farmers on whose land forest restoration will occur, is unquestionably complex (Table 3). To complicate implementation, the project is also reliant on local government, even local officials' personal interest and faith in the project, to be successful. A shift from a project management approach to a good governance approach is required.

Table 3. Map of stakeholders involved with a forest restoration program carried out on small landholdings in Protected Areas of Sustainable Development in Barra do Turvo, southeastern Brazil.

Stakeholder	Role	Scale of Action
Farmers	Providing areas on property for forest restoration; planting and maintenance of trees.	Local
NGO	Project concept, design and coordination; technical assistance.	International
São Paulo State Forest Foundation (FF)	Providing access to RDS and to farmers; project coordination.	State
BNDES (Brazilian Development Bank)	Providing project funding.	National
Forest restoration consulting company and the University of São Paulo	Project design and site assessments.	National; Atlantic Forest Biome
Local unions and farmer associations	Communication with farmers and responsibility for administrative concerns.	Regional

The NGO placed emphasis on conducting "participatory" workshops to design and implement forest restoration models with farmers. Counter intuitively, the degree of participation actually achieved through workshops and field visits may be more important to the NGO than to farmers, the majority of whom were not explicitly concerned with the project's long-term benefit to them when they agreed to participate in it. The NGO will rely on the representation of a participatory process, and of the project's "success," through reports and presentations to secure future funding from institutions that value participation, and the NGO has ultimate control over the "interpretation of events" [48].

In addressing trends in development project design and implementation, Mosse [48] describes the "mobilizing metaphors" of policy discourse, including "participation," "partnership," "governance," and "social capital." Because they can be interpreted broadly, these concepts feature centrally in project representation and in multi-stakeholder planning by serving to "conceal ideological differences, to allow compromise…and to multiply criteria of success within project systems" [48]. By adding a participatory component, not originally required by the forest restoration funded by BNDES, the NGO has rendered the project significantly more complex and must draw on existing development language and techniques for incorporating farmers into planning and implementation. In the present case, farmer participation was essentially limited to (1) choice to participate in a project that may bring future economic benefit, and some choice about where forest restoration will occur on their properties, and (2) choice of native species to plant. However, the forest restoration models themselves were designed apart from farmers, highlighting the fact that the entire structure and primary objectives of the project are necessarily non-participatory, requiring specialized technical knowledge. Farmers are invited to participate in very specific phases of the project, and although the design of the forest restoration models is meant to benefit smallholders, the primary objective is forest restoration.

This form of participatory engagement might be characterized by Mosse as a "commodity" of a project [48], holding an important symbolic position but effectively changing little in a project's central goals or technologies. Participants may come to appropriate these goals as their own in a process of "mirroring," whereby the "institutional needs of the project" become "built into community perspectives, making the project decisions appear perfectly participatory" [48]. In the present case, as the benefits of forest restoration and potential future benefits of economic native species are explained to farmers, farmers make decisions in line with the goals of the project. At the same time, details of project operation are modified to accommodate farmer ideas and needs, such as suggesting that they intercrop bananas and other annuals in initial stages of tree planting.

As discussed above, the project initially demonstrated low accountability [12] towards farmers by failing to adequately explain the purpose and intended outcomes of forest restoration on their lands, though this was significantly altered through subsequent field visits and workshops. The process of conducting workshops to better explain the project, to choose species in a participatory manner with farmers, and to provide training in area preparation and planting likely improved the trust between participants and the NGO. Through this process, the NGO both increased trust in the project [12] and its "downward accountability" towards a marginalized population, cited as a neglected component of multi-stakeholder implementation [49]. Not only must farmers demonstrate to NGOs and other authorities that they are capable of putting into practice project components, but these organizations must also show farmers that they are reliable and accessible.

Local civil society, such as farmer associations and cooperatives, can play a role in negotiating asymmetries between smallholders and more powerful actors, assuring just engagement of farmers by NGOs and improving farmer access to benefits brought by NGOs and government. NGOs themselves remain powerful actors in this asymmetry even as they may try to minimize it, at times unaware of how use of mobilizing metaphors such as participation in fact diminishes power sharing by setting the terms of smallholder engagement. In our case study, leaders of farmer associations were vocal in meetings and workshops in insisting that the NGO clarify intended benefits for farmers, and associations assumed responsibility for transferring money earned through the project from the NGO to farmers. Some of these leaders are individuals who share conservationist values and already have an interest in agroecology, and thus played key roles in influencing other farmers' perceptions of the project.

3.2.2. Problem of "community"

The concept of "community" acts as another kind of mobilizing metaphor, providing a site of intervention for projects. "Community-based natural resource management" requires a community, rather than individuals, to achieve equitable and sustainable resource management, though a discreet community upon which development can act may not always be present [21,22]. It is within the realm of community that the "environmental subject" emerges [50]; it is the unit upon which NGOs can act and for which they can most effectively attract funding. Communities, in turn, can reinforce this conceptualization as a space of intervention as a means to attract projects and attention from NGOs.

These environmental subjects, as "participants" in systems of environmental governance, come to perceive the environment as an object of governance by responding to incentives that necessitate sustainable management of natural resources [11,50]. In our example, smallholder farmers who have previous experience with conservation projects and exposure to conservation rhetoric are able to articulate perceptions of the environment using conservationist language and in some cases have altered their own perspectives on the environment and conservation as a result of this engagement. Here, project "success" is actually dependent on subject making [50], as the project will only accompany the farmers for the first few years and requires that farmers maintain interest in ensuring the success of tree growth and in pursuing avenues for commercialization of products derived from native species. Forest restoration success will also depend on farmers' increased valuation of environmentalist values of the landscape and decreased valuation of profits gained through cattle ranching or 'unproductive' farming.

Interviews at the household level demonstrated the diversity of opinion about the project, about conservation, and of production systems within each community. This variety reveals that in approaching members of the same "community," the NGO is basing its methodology on a simplified reality, seeing a homogenous community with common interests when it is in fact engaging individuals with different knowledge, experience, and opinions. Because they share similar production systems and cultural histories, Quilombola households seem to cohere as communities (as Quilombos) more neatly than family farmers in the other RDS, but conflict and diversity of opinion are still present within Quilombos. Intra-community conflict in all RDS include tension between those producing organically and those still using agrochemicals, and between ranchers who use fire to clear lands and their neighbors. Income disparity and conflict highlight the need to assure access by and opportunities for less powerful actors within communities when possible during the life of the project.

3.2.3. Trade-offs

As the political, social, and economic realities of this case study have demonstrated, the movement across scales in multilevel, multi-stakeholder development is a process of negotiating trade-offs. Development projects act as a social phenomenon that involves and affects various social actors or groups of actors, also called "strategic groups" [51], that interact and compete to capture the resources of a project. Thus, while projects involving diverse stakeholders should address the needs and priorities of every strategic group, strategies and outcomes fully satisfactory to each group cannot be expected. Rather, project management should focus on trade-offs acceptable to the parties. Trade-offs of this project include:

- Inability of every smallholder involved in the project to attend every meeting and workshop hosted by the NGO, due to lack of transportation or time. Thus, not all perspectives were taken into account, as the project in Barra do Turvo operated at the household rather than the community level.
- Design of forest restoration models on a university campus *versus* with farmer participants. However, the models were presented to participants in workshops, during which farmers were able to make recommendations for alterations. Species choices in the models were also primarily based on farmer suggestions.
- From the perspective of some farmers, losing productive space to forest restoration; from the perspective of the NGO, accepting less space per farmer property than preferred. These compromises were in some cases negotiated in the field during the prospecting phase, as farmers and project team members discussed current and potential future uses of pieces of land.
- Substitution of species more suitable to forest restoration for species with greater economic potential.
- Uncertainty of future legal situation conducive to commercialization of native species, but enough potential to design a project around the possibility.

Rather than "failures" or the less desirable alternatives to a win-win scenario, these trade-offs reflect realities of project implementation and of projects with conservation and development objectives. With improved project planning, such as better communication with farmers in initial stages, minimization of some trade-offs may be possible.

4. Conclusions

Large-scale forest restoration projects in protected areas, which involve small landholders and strive for both conservation and socio-economic development, are embedded in multi-scalar and complex social, legal and tenure contexts. Here, we have examined these contexts, including incentives for farmer participation, participatory project design and implementation, and questions of community and trade-offs. Studying the governance regime and relationships between the actors allows us to highlight the obstacles faced by the different stakeholders when designing and implementing a forest restoration project, as well as demonstrate the interdependence of the involved sectors.

Major barriers discussed include policy complexity and components of policy not necessarily aligned with the project objectives, and the uncertain evolution of legislation; administrative processes; the working unit (individual/household *versus* community) approached by the NGO which, if not properly defined, will lead to inappropriate proposals or inapplicable methodologies; and the lack of communication between parties.

We offer several recommendations that can improve the implementation of forest restoration initiatives involving smallholders, based in "good governance" that promotes "multilevel, nonhierarchical, information-rich, loose networks of institutions and actors" [11,52]. Good governance of forest restoration and conservation involving smallholders requires inclusion of and dialogue with farmers in all phases of the forest restoration process, as well as the need to adapt current legal instruments and incentives to this end. Recognition by institutional-scale governance bodies of the important role of local-level governance, and more serious incorporation of social science-based analyses prior to project implementation, will support the achievement of multiple goals, enhance power sharing, and reduce political asymmetry.

Increased attention to social analyses before and during project implementation aids in identifying relevant local, regional and even global policies [53]. Surveys and social evaluations at the outset of projects, and thorough investigations of historical, cultural and economic backgrounds, also significantly contribute to better understandings of the strategies of the participants, allow projects to appropriately adapt, and increase the acceptability of projects by smallholders. Pre-implementation social analysis also improve institutions' (NGO, government, university) understanding of local farmers' relationship with their landscape, and how their sense of place is formed by daily interactions with it. Improving participatory techniques, working from local relationships with landscape, and establishing a relationship of trust through frequent contact can minimize trade-offs and ensure participation throughout the project. Civil society can play a role in negotiating this trust, in improving smallholder access, and in promoting openness and accountability.

Finally, we stressed the "flexibility" and interdependence of the concerned institutions. Because institutions must deal with uncertainty in environmental projects [6,12], they should be ready to adapt and adjust to the reality of the field, to small farmers' needs, and to environmental and legal variability. Forest restoration projects must be concerned with both conservation and livelihoods, as recognized by RNC, and can provide alternatives to conventional forest restoration that not only increase the ecological complexity of the system to be restored, but also transform the socioeconomic landscape. Forest restoration projects must compensate the loss of arable lands and offer economic incentives, such as contracting farmers for planting and including crops and exotic species in agro-successional models that will evolve into production areas of timber and non-timber forest products that can be sustainably managed. In the Atlantic Forest, management of economically interesting species such as *E. edulis* can address both forest restoration and development goals, with the objective of avoiding little success in either, or significantly more success in one realm than the other.

Acknowledgments

The authors wish to thank the Laboratory of Tropical Silviculture (LASTROP) and the Laboratory of Ecology and Forest Restoration (LERF) of the Escola Superior de Agricultura "Luiz de Queiroz" for logistical support, the Yale MacMillan Center Fox International Fellowship, and the described NGO for the opportunity to accompany the project described here. We are grateful to our colleagues in the field for their insights and enthusiasm, to the farmers of Barra do Turvo for participating in this research, and to James Aronson and two anonymous reviewers for their valuable feedback on drafts of the article.

Author Contributions

Alaine Ball and Alice Gouzerh carried out the fieldwork described in the article, with both authors present during NGO-organized trips. A. Gouzerh conducted separate, structured interviews with farmers without the presence of the NGO. Pedro Brancalion advised the research of both Alice Gouzerh and Alaine Ball. All authors contributed to the writing of the article and the analysis presented here.

Conflicts of Interest

The authors declare no conflict of interest.

References

1. Millennium Ecosystem Assessment. World Resources Institute: Washington, DC, USA, 2005. Available online: www.millenniumassessment.org (accessed on 29 December 2013).
2. Bulkeley, H. Reconfiguring environmental governance: Towards a politics of scales and networks. *Polit. Geogr.* **2005**, *24*, 875–902.
3. World Bank. *Governance and Development*; World Bank Publications: Washington, DC, USA, 1992. Available online: http://elibrary.worldbank.org/doi/book/10.1596/0-8213-2094-7 (accessed on 29 December 2013).
4. Doornbos, M. Good Governance: The metamorphosis of a policy metaphor. *J. Int. Aff.* **2003**, *57*, 2–17.
5. Jose, J. Reframing the 'governance' story. *Aust. J. Polit. Sci.* **2007**, *42*, 455–470.
6. Batterbury, S.P.J.; Fernando, J.L. Rescaling governance and the impacts of political and environmental decentralization: An introduction. *World Dev.* **2006**, *34*, 1851–1863.
7. UNDP. Governance for sustainable human development. United Nations Development Programme, New York, NY, USA, 1997. Available online: http://mirror.undp.org/magnet/policy/ (accessed on 13 December 2013).
8. Leftwich, A. Governance, democracy and development in the third world. *Third World Q.* **1993**, *14*, 605–624.
9. Sunderlin, W.D.; Angelsen, A.; Belcher, B.; Burgers, P.; Nasi, R.; Santoso, L.; Wunder, S. Livelihoods, forests, and conservation in developing countries: An overview. *World Dev.* **2005**, *33*, 1383–1402.
10. Liverman, D. Who governs, at what scale, and at what price? Geography, environmental governance, and the commodification of nature. *Ann. Assoc. Am. Geogr.* **2004**, *94*, 734–738.
11. Lemos, M.C.; Agrawal, A. Environmental governance. *Ann. Rev. Environ. Resour.* **2006**, *31*, 297–325.

12. Lebel, L.; Anderies, J.M.; Campbell, B.; Folke, C.; Hatfield-Dodds, S.; Hughes, S.T.P.; Wilso, J. Governance and the capacity to manage resilience in regional social-ecological systems. *Ecol. Soc.* **2006**, *11*, 19.

13. Folke, C.; Hahn, T.; Olson, P.; Norberg, J. Adaptive governance of social-ecological systems. *Ann. Rev. Environ. Resour.* **2005**, *30*, 441–473.

14. Note 1 Used here to refer to the capacity of social and ecological systems to experience change and disturbance without fundamental shifts in the processes that control the system. Self-organization, learning, and adaptation are key elements in a resilient system (Resilience Alliance, 2002. Available online: http://www.resalliance.org/index.php/resilience (accessed on 4 March 2014.

15. Defining Civil Society. World Bank, 2013. Available online: http://go.worldbank.org/ 4CE7W046K0 (accessed on 12 December 2013).

16. Refers to the wide array of non-governmental and not-for-profit organizations that have a presence in public life, expressing the interests and values of their members and other citizens, based on ethical, cultural, political, scientific, religious or philanthropic considerations [14].

17. Bebbington, A.; Thiele, G.; Davies, P.; Prager, M.; Riveros, H. *Non-Governmental Organizations and the State in Latin America: Rethinking roles in Sustainable Agricultural Development*; Overseas Development Institute, Routledge: London, UK, 1993.

18. Roy, I. Civil society and good governance: (Re-) conceptualizing the interface. *World Dev.* **2008**, *36*, 677–705.

19. Paavola, J. Institutions and environmental governance: A reconceptualization. *Ecol. Econ.* **2007**, *63*, 93–103.

20. While institutions can be broadly conceived of as all "sets of formal and informal rules and norms that shape interactions of humans with others and nature" [21], here we limit our use of 'institutional' to refer to government agencies, civil society organizations, and universities, in contrast with the 'local' perspective. For a detailed discussion of institutions and environmental justice, see Paavola 2007 [19].

21. Agrawal, A.; Gibson, C.C. Enchantment and disenchantment: The role of community in natural resource conservation. *World Dev.* **1999**, *27*, 629–649.

22. Rose, N. Community. In *Powers of Freedom: Reframing Political Thought*; Cambridge University Press: Cambridge, UK, 1999; pp. 167–196.

23. McShane, T.O.; Hirsch, P.D.; Trung, T.C.; Songorwa, A.N.; Kinzig, A.; Monteferri, B.; Welch-Devine, M.; Brosius, J.P.; Coppolillo, P.; O'Connor, S. Hard choices: Making trade-offs between biodiversity conservation and human well-being. *Biol. Conserv.* **2011**, *104*, 966–972.

24. Society for Ecological Restoration International Science and Policy Working Group. *The SER International Primer on Ecological Restoration*; Society for Ecological Restoration International: Washington, DC, USA, 2004. Available online: www.ser.org (accessed on 23 August 2013).

25. Vogler, J. Taking institutions seriously: How regime analysis can be relevant to multilevel environmental governance. *Glob. Environ. Polit.* **2003**, *3*, 25–39.

26. *Restoration Ecology: A Synthetic Approach to Ecological Research*; Jordan, W.R., Gilpin, M.E., Aber, J.D., Eds.; Cambridge University Press: Cambridge, UK, 1990.

27. Chapin, M. A Challenge to conservationists. *World Watch Magazine.* **2004**, *17*, 17-31.

28. Billé, R.; Chabason, L. La conservation de la nature : Origines et controverses. In *Regards sur la Terre*; Presses de Sciences Po: Paris, France, 2007; pp. 113–130.

29. *Restoring Natural Capital: Science, Business and Practice.* The science and practice of ecological restoration series. Aronson, J., Milton, S.J., Blignaut, J.N., Eds.; Island Press: Washington, DC, USA, 2007.

30. Aronson, J.; Alexander, S. Restoration ecology, landscape ecology and conservation biology: Three sisters on the path to sustainability. *Natureza Conservação* **2014**, in press.

31. Blignaut, J.N.; Aronson, J.; de Groot, R.S. Restoration of natural capital: A key strategy on the path to sustainability. *Ecol. Eng.* **2013**, in press.

32. RNC Alliance, http://www.rncalliance.org/ (accessed on 25 August 2013).

33. Brancalion, P.H.S.; Viani, R.A.G.; Strassburg, B.B.N.; Rodrigues, R.R. Finding the money for tropical forest restoration. *Unasylva* **2012**, *63*, 25–34.

34. Blanc-Pamard, C.; Fauroux, E. L'illusion participative: Exemples ouest-malgaches. *Autrepart* **2004**, *31*, 3–19.

35. Porter-Bolland, L.; Ellis, E.A.; Guariguata, M.R.; Ruiz-Mallén, I.; Negrete-Yankelevich, S.; Reyes-García, V. Community managed forests and forest protected areas: An assessment of their conservation effectiveness across the tropics. *For. Ecol. Manag.* **2012**, *268*, 6–17.

36. Laurance, W.F. Conserving the hottest of the hotspots. *Biol. Conserv.* **2009**, *142*, 1137.

37. Plummer, R.; Fitzgibbon, J. Co-management of natural resources: A proposed framework. *Environ. Manag.* **2004**, *33*, 876–885.

38. Ribeiro, M.C.; Metzger, J.P.; Martensen, A.C.; Ponzoni, F.J.; Hirota, M.M. The Brazilian Atlantic Forest: How much is left, and how is the remaining forest distributed? Implications for conservation. *Biol. Conserv.* **2009**, *142*, 1141–1153.

39. Mapa do Mosaico do Jacupiranga, Instituto Socioambiental, 2008. http://site-antigo.socioambiental.org/nsa/detalhe?id=2614 (accessed on 23 September 2013).

40. Caiçaras are descendants of white Portuguese and indigenous groups forming coastal communities of traditional fishermen and farmers that span the southeastern Brazilian coast; and Quilombolas are descendants of escaped slaves who depend on traditional swidden agriculture.

41. Brancalion, P.H.S.; Vidal, E.; Lavorenti, N.A.; Batista, J.L.F.; Rodrigues, R.R. Soil-mediated effects on potential *Euterpe edulis* (*Arecaceae*) fruit and palm heart sustainable management in the Brazilian Atlantic Forest. *For. Ecol. Manag.* **2012**, *284*, 78–85.

42. The MOJAC (created in 2008) is comprised of 14 protected areas, including the RDS of the project, spans 234,885 ha, and is home to over 8000 people.

43. Mazoyer, M.; Roudart, L. *Histoire des agricultures du monde: Du néolithique à la crise contemporaine*; Éditions du Seuil: Paris, France, 1987.

44. SMA 16/1994 defines management regulations for E. edulis heart of palm, but this instrument is little used for legal exploitation of heart of palm due to the bureaucracy and investment required to correctly implement it.

45. Vieira, D.L.M.; Holl, K.D.; Peneireiro, F.M. Agro-successional restoration as a strategy to facilitate tropical forest recovery. *Restor. Ecol.* **2009**, *17*, 451–459.

46. Brancalion, P.H.S.; Cardozo, I.V.; Camatta, A.; Aronson, J.; Rodrigues, R.R. Cultural ecosystem services and popular perceptions of the benefits of an ecological restoration project in the Brazilian Atlantic Forest. *Restor. Ecol.* **2014**, *22*, 65–71.

47. Aronson, J.; Durigan, G.; Brancalion, P.H.S. Conceitos e definições correlatos à ciência e à prática da restauração ecológica. *IF Série Registros* **2011**, *44*, 1–38.

48. Mosse, D. Is good policy unimplementable? Reflections on the ethnography of aid policy and practice. *Dev. Chang.* **2004**, *35*, 639–671.

49. Ribot, J.C. *African Decentralization: Local Actors, Powers and Accountability*; United Nations Research Institute for Social Development: Geneva, Switzerland, 2002; p. 8.

50. Agrawal, A. Environmentality: Community, intimate government, and the making of environmental subjects in Kumaon, India 1. *Curr. Anthr.* **2005**, *46*, 161–190.

51. Bierschenk, T. Development projects as arenas of negotiation for strategic groups: A case study from Bénin. *Sociol. Rural* **1988**, *28*, 146–160.

52. Haas, P. Addressing the global governance deficit. *Glob. Environ. Polit.* **2004**, *4*, 1–15.

53. Jorgensen, D.; Nilsson, C.; Hof, A.R.; Hasselquist, E.M.; Baker, S.; Chapin III, F.S.; Eckerberg, K.; Hjälten, J.; Polvi, L.; Meyerson, L.A. Policy language in restoration ecology. *Restor. Ecol.* **2014**, *22*, 1–4.

A Comparison of Governance Challenges in Forest Restoration in Paraguay's Privately-Owned Forests and Madagascar's Co-managed State Forests

Stephanie Mansourian, Lucy Aquino, Thomas K. Erdmann and Francisco Pereira

Abstract: Governance of forest restoration is significantly impacted by who are the owners of and rights holders to the forest. We review two cases, Paraguay's Atlantic forest and Madagascar's forests and shrublands, where forest restoration is a priority and where forest ownership and rights are having direct repercussions on forest restoration. In Paraguay where a large proportion of forests are in the hands of private landowners, specific legislation, government incentives, costs and benefits of forest restoration, and the role of international markets for commodities are all key factors, among others, that influence the choice of private landowners to engage or not in forest restoration. On the other hand, in Madagascar's co-managed state forests, while some similar challenges exist with forest restoration, such as the pressures from international markets, other specific challenges can be identified notably the likely long term impact of investment in forest restoration on land rights, traditional authority, and direct links to elements of human wellbeing. In this paper, we explore and contrast how these different drivers and pressures affect the restoration of forests under these two different property regimes.

Reprinted from *Forests*. Cite as: Mansourian, S.; Aquino, L.; Erdmann, T.K.; Pereira, F. A Comparison of Governance Challenges in Forest Restoration in Paraguay's Privately-Owned Forests and Madagascar's Co-managed State Forests. *Forests* **2014**, *5*, 763-783.

1. Introduction

Forest restoration is increasingly being seen as an option to combat the degradation, loss and fragmentation of tropical forests. In the Atlantic forest of Paraguay and the moist forests of Madagascar, reforestation and forest restoration have been used as tools to counter forest loss. While reforestation refers to the return of trees to a previously forested land, it is frequently associated with the use of exotic species (e.g., [1,2]). On the other hand, forest restoration aims to recover most or all of a reference ecosystem. The Society for Ecological Restoration defines restoration as "the process of assisting the recovery of an ecosystem that has been degraded, damaged, or destroyed" [3]. Increasingly, many restoration projects focus on restoring ecosystem services [4], which may not always correspond to reference ecosystems or lead to improvements in biodiversity. Yet, natural forests composed of indigenous species are more adapted to local climatic conditions, provide local animal species with their native habitat, are more resilient and have traditionally been used by local inhabitants as a source of numerous products and services (e.g., [2,5–7]). The success or failure of forest restoration is frequently associated with underlying governance challenges, which are all too often overlooked.

Governance of forests (and natural resources more generally) encompasses a range of dimensions, notably related to who takes decisions, how these are taken and what mechanisms exist for effective decision-making related to natural resources (e.g., [8]). In small areas with clear property rights and a single landowner (state or other), decisions are somewhat easier to take although they may be complicated by underlying conflicting land claims (e.g., [5,7]). However, in larger areas (landscapes) where different land owners and users are involved, governance issues become more complex (e.g., [9,10]).

Legal forest ownership can be categorized as public or private, with community ownership and traditional ownership straddling these classifications. Management of forests can also be further sub-categorized as community, private, government, or co-managed [5,11]. Globally approximately 80% of forests are publicly owned, while 17.8% are privately owned and 2.2% classified as under "other" ownership [12]. These figures hide regional differences and conflicting claims over recognition of land and forest rights [13]. In 2002, a review by White and Martin [14] provided the following figures: 77% owned and administered by governments, 4% reserved for communities, 7% owned by local communities, and approximately 12% owned by individuals. In 2008, a further review [13] demonstrated that for 25 of the top 30 forested countries (covering 80% of the global forest estate) there was a reduction in state-owned forests (to 74%) with the remainder shared between communities, individuals and firms. Furthermore, management responsibilities are also slightly different with 80% of forests managed by the state, while private corporations and institutions manage 10% of the world's forests and communities manage 7% [12]. A general trend towards decentralization of forest management can be seen globally [15] which may or may not facilitate the claims of forest-dependent communities [13]. Unclear tenure appears to be an important cause of failure in managing (and restoring) forests [16]. We explore how different governance challenges appear exacerbated or complicated under different forest tenure arrangements leading to more or less effective forest restoration in Paraguay's Atlantic forest and Madagascar's forest and shrublands ecoregion.

2. Experimental Section

2.1. Methodology

The objective of this work is to compare and contrast the different factors influencing the success (or failure) of forest restoration under two different property regimes in two of the world's biodiversity hotspots. In Paraguay, the focus is on private forests and in Madagascar on forests that are co-managed by the State and local community associations. Furthermore, in both cases, forest restoration is undertaken as one of the components of forest management (rather than a standalone priority). We compared the importance and threats to forests in Madagascar and Paraguay in order to understand the emergence and role of forest restoration. In particular, we looked at recent (twentieth and twenty-first century) historical changes in forest cover, land use, and relevant legislation (specifically, incentives and policies or policy frameworks related to forest management, use and restoration).

Our approach relied on an extensive literature review. A number of interviews were undertaken either by phone, Skype or face to face in Spanish and French to corroborate some of our findings and to add to our dataset. Interviewees were selected because of their direct experience in implementing forest restoration activities and/or forest co-management contracts (Madagascar) or because they were landowners undertaking forest restoration (Paraguay). Interview questions can be found in Appendix 1. This paper also builds on direct field work by three of the authors.

2.2. Framing Governance of Forest Restoration

The success or effectiveness of forest restoration is influenced by a range of factors, including policies, incentives, land tenure, and markets, to cite just a few. It is also influenced by actors at all levels, from local to international. Several environmental governance frameworks exist which can be adapted to forest restoration. Lemos and Agrawal [8], for instance, highlight that environmental governance equates to interventions aiming at "changes in environment-related incentives, knowledge, institutions, decision making, and behaviors". They also identify the importance of the mechanisms, processes, regulations and organizations in governance to influence environmental outcomes. For Kishor and Rosenbaum [16] forest governance relates to "the norms, processes, instruments, people, and organizations that control how people interact with forests." Authority, power and capacity are three key dimensions considered by USAID [10] for effective natural resource governance. Davis *et al.* [17] refer to "actors" (including people and institutions), "rules" (including policies and laws) and "practices", as three essential components of forest governance. In this paper, we use a similar framework (see Figure 1) adapted from Mansourian and Oviedo [18] to explore, compare and contrast the governance factors that influence forest restoration in Madagascar and Paraguay.

Figure 1. Framework to Assess the Governance of Forest Restoration (adapted from [18]).

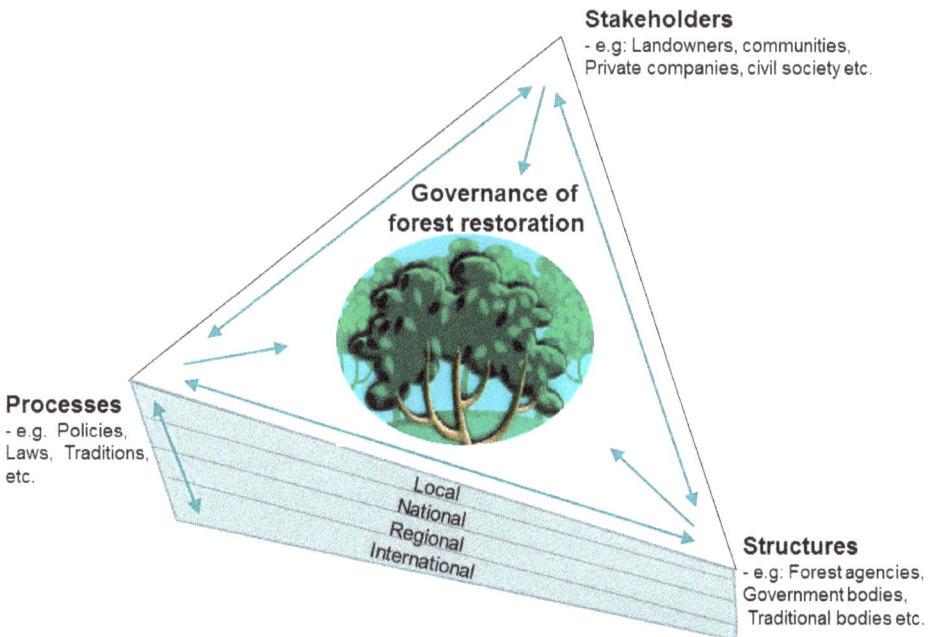

The framework proposed considers three essential factors when exploring the governance of natural resources, and in this particular case, the governance of forest restoration, these are: processes, stakeholders and institutions. "Processes" include policies, laws, strategies and all relevant rules. For example, in the context of forest restoration, processes might include laws related to land tenure or subsidies for planting different species. "Structures" in this framework include different agencies and other relevant bodies that help to organize stakeholders. In the context of forest restoration, this could be research bodies or implementing agencies, for example, local community groups or at a different scale, the national forest service. As to "stakeholders", in the context of forest restoration they may be local communities, individual landowners, private companies, and the government, amongst others. While the three dimensions impact on forest governance, they also inter-relate (see arrows in the diagram) in ways that may either complicate or simplify forest governance. For example, if representative bodies exist (under the "structures" dimension), communities (under the "stakeholders" dimension) may voice their needs and aspirations more effectively, thus leading (potentially) to these needs being better integrated into forest restoration, and overall, to better forest governance. All three dimensions of the framework provide essential foundations for successful restoration. Furthermore, they can be found at different scales, from local to international (also see for e.g., [19]). It can be argued that effective governance that supports forest restoration requires that all three dimensions be functioning optimally and also that the interactions between them function effectively. Although the emphasis in this paper is on the "processes" dimension, the other two dimensions are also considered.

2.3. Study Sites

Both Paraguay and Madagascar harbor two of the world's biodiversity hotspots as described by Myers *et al.* [20] and priority global ecoregions [21]: the Atlantic forest ecoregion (Argentina, Brazil and Paraguay) and Madagascar's forest and shrublands ecoregion. The specific zones explored in this paper are the Oriental Region of Paraguay (north and southeastern part of the country) and the moist tropical forest zone of Madagascar (eastern half of the island) (see maps in Figure 2).

Economically, Paraguay is classified by UNDP's Human Development Index as medium development (rated 111th out of 186 countries) while Madagascar is classified as a low development country rated 151st [23]. Both countries have suffered from high rates of deforestation in the last decade, with Paraguay's deforestation rate averaging 16% between 1999 and 2010, while Madagascar's was at a rate of 8% during the same period [23]. In terms of forest cover, Paraguay is classified as 44% forest while Madagascar's land cover is 21.6% forest [23]. High deforestation rates have been fuelled in both countries by the economy increasingly relying on forest exploitation and conversion: in the case of Paraguay for energy, commercial agricultural and livestock [24–26], and in Madagascar both for energy and subsistence agriculture [27] (see Table 1).

Figure 2. Forest cover of Paraguay and Madagascar.

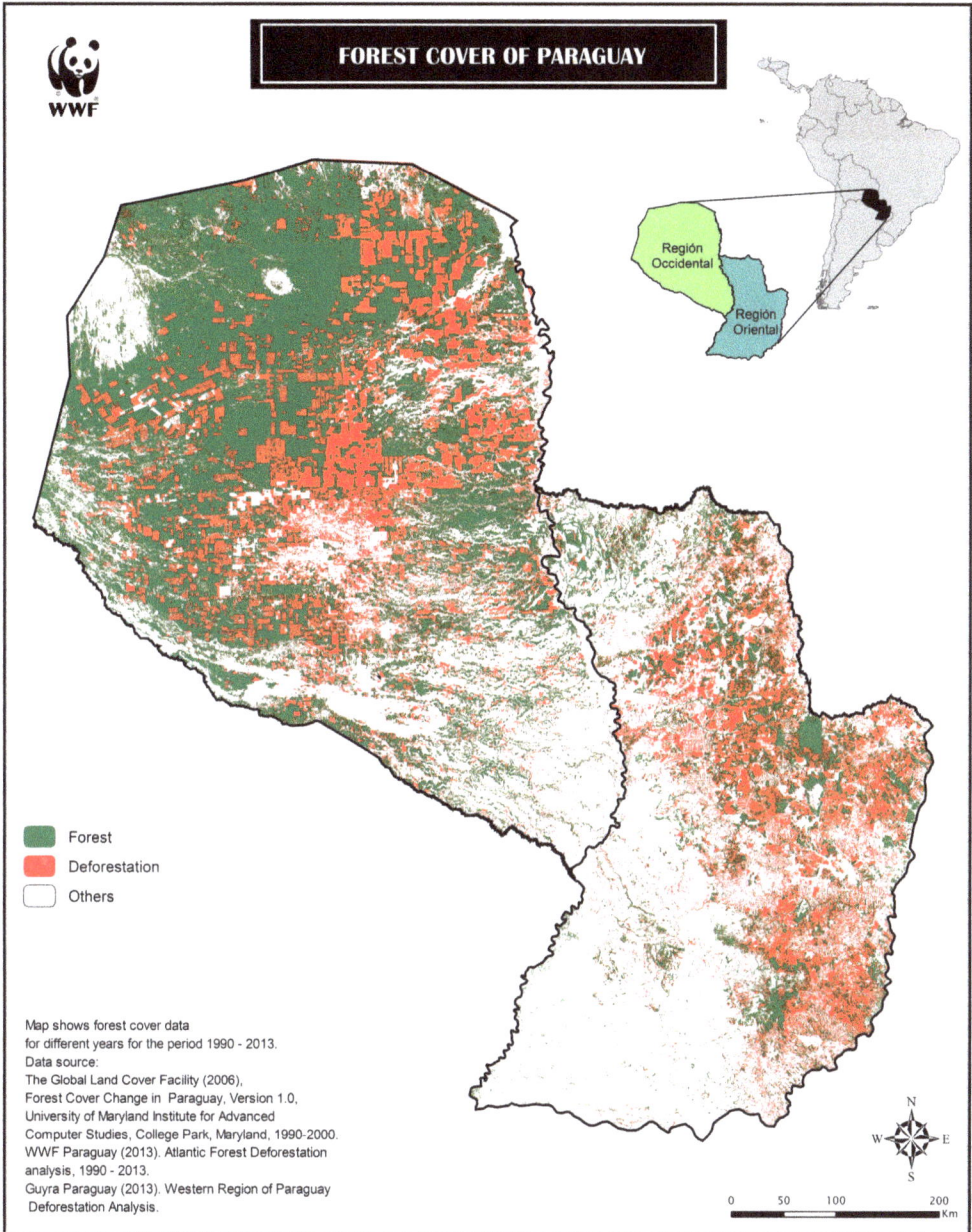

Figure 2. *Cont.*

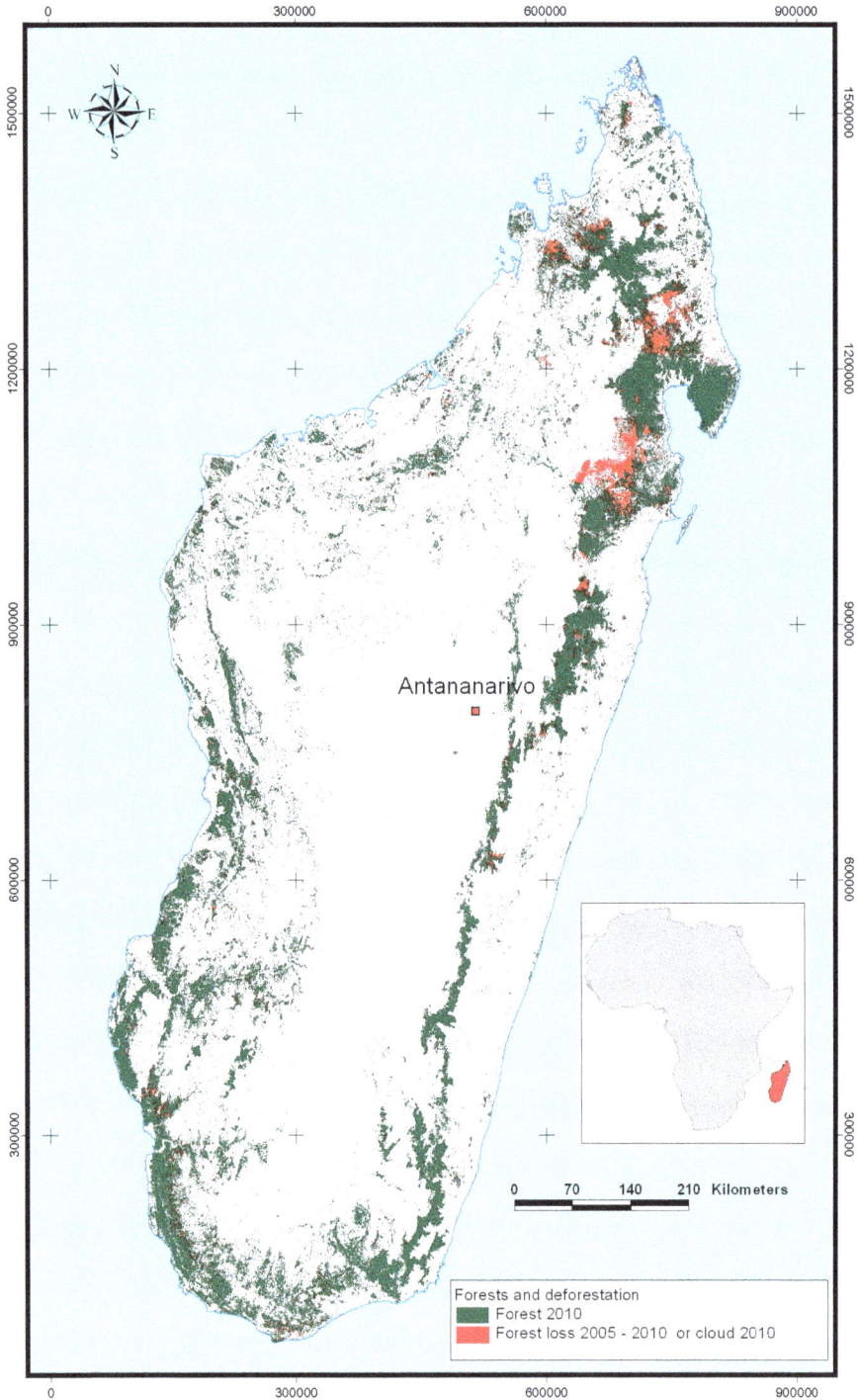

Source: Office National pour l'Environnement, Madagascar [22].

Table 1. Forest Cover in Madagascar and Paraguay.

Country	Primary Forest		Other Naturally Regenerated Forest		Planted Forest		Annual Rate of Change in Forest (Total) 2005–2010	
	1000 ha	% of forest	1000 ha	% of forest	1000 ha	% of forest	1000 ha/year	%
Paraguay	1850	11	15684	89	48	n.s.	−179	−0.99
Madagascar	3036	24	9102	73	415	3	−57	−0.45

Source: [12].

Other sources provide higher rates of deforestation for both Madagascar (for e.g., [28]) and Paraguay (for e.g., [25]).

2.4. Overview of Forests in Each Ecoregion

Madagascar's forests were estimated in 2005 by FAO [12] to be predominantly under public ownership (see below), while Paraguay was officially nearly two-thirds (61%) under private ownership although the actual figure is considerably higher, estimated at over 90% (see for e.g., [29]). Of the 98% under public ownership in Madagascar, management rights for 2% have been devolved to communities. In addition, in Madagascar, of the 2% under private ownership, the majority (92%) is owned by individuals, with 8% owned by local, indigenous and tribal communities [12] (see Table 2).

Table 2. Forest ownership in Madagascar and Paraguay.

Country	Ownership (2005)	
	Public	Private
Madagascar	98	2
Paraguay	39	61

Source: [12].

2.4.1. Paraguay's Atlantic Rainforest

The Atlantic Forest ecoregion complex extends across the three South American countries of Paraguay, Argentina and Brazil. Only an estimated 11.7% of the Atlantic forest's original area remains [30]. It consists of 15 distinct sub-ecoregions, with the one in Paraguay classified as Upper Paraná Atlantic Forest. The Atlantic forest is characterized by semi-humid forest with annual rainfall lower than 1700 mm and humid forest with rainfall between 1900–2000 mm. Paraguay's Upper Parana Atlantic Forest is home to an extremely varied flora including vascular plants, pteridophytes and bryophytes [31]. The forest is heavily fragmented, and restoring connectivity among forest fragments is a priority to improve functionality [30].

More than 97% of Paraguay's over six million inhabitants live in the Oriental region of the country, the area once dominated by the Atlantic Forest. While the Oriental Region makes up just 39% of the total land area of the country, the population density is disproportionately greater than in the rest of the country [32]. Massive loss and degradation of the forest can be attributed to demand for fuelwood, and agricultural and livestock development.

Until the mid-1960s Paraguay's eastern region remained largely untouched. Severe deforestation started in the 1960s and continued increasing in the 1970s with the so-called "green revolution", for the development of agriculture (cotton and soy) and the conversion of forest to exotic pasture for cattle ranching [33]. This was followed by extensive soy cultivation (with Paraguay being one of the world's leading soy exporters) [34].

2.4.2. Madagascar's Forests and Shrublands Ecoregion

Madagascar's forests and shrublands ecoregion is located along the eastern escarpment and coastal plain of the island. The ecoregion includes moist forests across an altitudinal range from sea level up to over 2000 meters (it includes: lowland rain forest (0 to 800 m), moist montane forest (800 to 1300 m) and sclerophyllous montane forest (1300 to roughly 2300 m) [35]. These forests have long been recognized as an important center of endemism and diversity with hundreds of species of vertebrates and thousands of species of plants being strictly endemic to this ecoregion. All five families of endemic Malagasy primates can be found here, as can seven endemic genera of Rodentia, six endemic genera of Carnivora, and several species of Chiroptera [36].

Despite its importance, much of this habitat has been removed or fragmented (with an estimated 830 million ha being fragmented [27]). The predominant cause of deforestation is the local tradition of "tavy" or slash and burn agriculture (mainly for rainfed rice and cassava cultivation) which accounts for 80% of deforestation [27], although some [37] trace the process of deforestation back to the French occupation, particularly because of their logging concessions and cash crop plantations. Yet the Malagasy population is to a large extent (up to 70% according to [27]) dependent on forests—both plantations and natural forests—for fuelwood and construction materials [38].

Plantations are dominated by exotic species in Madagascar, with pines, eucalypts, and wattles among the more common species; shade, ornamental, and fruit trees are also planted around settlements. These have been promoted through government efforts to reforest notably via state-owned plantations and land tenure incentives [39,40].

3. Results and Discussion

The weight or importance attributable to different governance factors varies under different ownership and management systems for forests in the two case study countries and regions. Here we describe the key aspects of forest ownership and management in both countries and regions.

3.1. Ownership and Management of Forests

3.1.1. Madagascar

While much of Madagascar's land is under customary tenure arrangements, without deeds, titles, or cadastres [40], in actual fact customary tenure is rarely legally recognized [38]. Lack of clear tenure has been identified as one of the underlying causes of deforestation in Madagascar [41]. For example in the "payment for ecosystem services" project in Mantadia, the challenge identified by Wendland *et al.* [42] has been dealing with property rights since although most of the land is state-owned, individual and communal entitlements exist leading to conflicting land claims. Since 2005 a project to reform land tenure (Programme National Foncier—PNF or "national tenure program") has been in place, which could improve recognition of customary rights, although in the meantime the coup d'état has severely slowed such reforms. In addition, the program focuses on improving titling for agricultural land rather than forests [43].

Madagascar's attempts at improving community engagement in forest management started in 1996 with the creation of the GELOSE (Gestion Locale Sécurisée or "secure local management") law, which allows for the devolution of management rights of natural resources to rural communities. In 2001, a further decree established the Gestion Contractualisée des Forêts (GCF or "contractual management of forests") law, which defined the details of the contracts to transfer forest management rights and streamlined the GELOSE process [44].

Concretely, in co-managed forests three zones can usually be found: one under conservation where extraction is banned, one under sustainable management of resources for local community use, and a commercial zone. In some cases, a fourth restoration zone is designated. Groups of communities have to organize themselves in COBA (communautés de base or "local communities") associations to sign official management transfer contracts. These are initially valid for three years but can then be renewed. Contracts are negotiated between the central government, the commune and local communities [45]. To this day all the management contracts have been mediated by external agencies, such as international NGOs.

An estimated 750 co-management contracts have been signed since the start of this process for an area of more than 1.2 million hectares [46]. Furthermore, there has been a recognized growth in the capacity of communities to manage their forests more generally and to understand the stakes. According to Randrianarisoa *et al.* [47], in some areas under co-management deforestation has also been reduced although this has certainly not been the case across all areas under co-management. Recently, broader governance issues affecting the country have also had repercussions on environmental governance, including a marked increase in illegal logging [48].

3.1.2. Paraguay

In Paraguay in contrast, the majority of the nation's forests are privately owned (61% according to FAO data [12] but closer to 90% according to other sources [29]. Land tenure remains one of the major causes of conflict in Paraguay.

Management of forests has been regulated by a number of laws, although in practice, there has been little enforcement [49]. Indeed, as highlighted by Contreras-Hermosilla [50] over-regulation characterizes much of Latin America's forestry sector, with in actual fact poor environmental outcomes. Ultimately, overly complex and multiple laws in the forest sector tend to lead to poor enforcement and alternative uses of land (notably for agricultural commodities) where legislation is more straightforward [51].

First and foremost, Paraguay still needs to develop an agrarian reform to distribute land equitably and implement land use planning. While multiple proposals have been made for reform, full reform (in 2001 the Agrarian Statute was amended to remove the consideration that forest lands are unproductive areas) has still not happened and landless peasants continue claiming land for agriculture and demanding an agrarian reform. Paraguay's Agrarian Statute of 1963 provided a perverse incentive for forest owners to clear land and put it under "productive use" lest this "unproductive land" be claimed by small farmers.

Since 1973, the Forest Law (Forestry Law 422/73) states that 25% of all land should remain under forest cover. It also establishes fiscal incentives for reforestation [52]. However, loopholes exist in this law whereby by transferring the 25% to other owners, this area could be further cleared by 75%. The result is that in the Eastern Region of Paraguay, forest cover is below 10% on private land [53].

To help promote recovery of the Atlantic forest in compliance with Article 42 of Forest Law 422/73, the Conformance with Forest Law (CFL) program was recently created. This program constitutes a legal tool and market mechanism to help forest owners meet the required minimum 25% forest cover. For each property equal to or greater than 20 ha, the shortfalls or profits are calculated using satellite images (LANDSAT 5 TM and C-BERS 2B, provided for free by the Instituto de Pesquisas Espaciais (INPE or the "institute of space research") of Brazil). Properties with an environmental shortfall are defined as those with the partial or complete absence of the 25% forest reserve cover and/or protective gallery forest along watercourses. In this way, environmental "profits" and "shortfalls" are quantified in terms of hectares of forest per district and per landowner. However, CFL can only be applied in areas where an up-to-date official register of land tenure is available. Currently only two political departments—Itapúa and Alto Paraná—fulfill this condition.

A further law (Law 4241—Law on Protected Forests along Watercourses) was passed to encourage restoration of riverine forest. This law is promoted via the provision of saplings from tree nurseries, although quantities were limited and only five different species were provided as of 2004.

A law to promote reforestation (Ley No. 536 "Law to Promote Forestation and Reforestation") was enacted in 1995 establishing economic incentives and subsidies for forestry plantations with the government providing up to 75% of the direct costs of reforestation. However, the government has been unable to find the financial resources to maintain this program. In 2004, reforestation achievements were estimated to total approximately 40,000 hectares [53].

Sustainable forest management has been very limited in Paraguay (due to the high price of the certification process and the complicated bureaucracy involved) and currently there is only one Private Reserve (Ypeti) in the Atlantic Forest with FSC (forest stewardship council) certification. Illegal and legal forest management efforts have been very hard to tell apart in Paraguay and the government has failed to invest in supporting the forestry sector and protect landowners from unfair competition from illegal logging. To this day, the INFONA (The National Forestry Institute) does not have a sound system to record forest management operations [54]. As a result, private forest lands (and even protected areas) suffer severe degradation, and illegal conversion of forest is frequent, starting with selective logging induced by forest fire. Once the forest has been degraded, landowners disseminate exotic grass seeds that quickly dominate the area and the land is declared by the authorities as a cattle ranching area. Today this constitutes the most common process of deforestation in the Oriental Region of Paraguay.

A moratorium on deforestation was established since December 2004 in the Oriental Region of Paraguay, prohibiting clear cuts [55] which has now been renewed twice already until 2018. Even though this legislation helps to decrease the deforestation rate in the Oriental Region, degradation and illegal conversion continue by landowners, supported by the lack of will and even cases of corruption in the government.

The government, with the support of NGOs is now developing a protocol for Law 3001 (a "Payment for Environmental Services" law) that will pay landowners who conserve and protect their forest (additional to the required 25%). The implementation of these payments is now being refined and may become a good incentive for landowners to engage in restoration on their land.

3.2. Selection of Restoration/Reforestation Projects

Through interviews and a literature review we identified some recent projects in Paraguay and Madagascar that included restoration or reforestation as one objective (see Table 3). In Madagascar, many of the projects involve payments for an ecosystem service such as biodiversity conservation or carbon sequestration (through REDD or another carbon-related mechanism). In Paraguay the protocol for payments for ecosystem services is still at an early stage and does not yet include restoration but rather avoiding deforestation.

A distinction is made for these projects between ownership and management, which is particularly relevant to Madagascar where co-management is in place on public forest land. Through interviews, we attempted to specify the principal aims of forest restoration in the two countries, the way it was undertaken and the challenges involved. Three main categories of aims for restoration (or reforestation *sensu lato*) were identified: ecological, socio-political and financial. Table 4 below summarizes our findings.

3.3. Discussion

It appears that success in ecological restoration remains limited in both countries given the continued rates of forest loss. In the face of this result, we explore the associated governance challenges. Taking the three dimensions of our forest governance framework, the processes dimension appears to be the most important in both country contexts. While the discussion focuses on this dimension, we also discuss the role of key stakeholder groups in the context of governance of forest restoration in Paraguay and Madagascar.

3.3.1. Processes

The main factors that appear to emerge from this case study comparison as being critical to forest restoration are related to tenure, management rights and incentives—all three falling under the "processes" dimension of our framework.

The role of land tenure within the framework of forest restoration and the pressure on forests in the context of tenure security or insecurity [51,56,57] is further highlighted through the above comparisons and study. In Paraguay, despite specific laws to promote reforestation and to stop deforestation in the Oriental Region, the country continues to face a high deforestation rate in both the Occidental and Oriental Regions [58] and limited success with restoration. This can be attributed to limited reach and capacity of the government to apply and enforce laws, many of which actually over-burden the sector. Wright *et al.* [59] also associate the high deforestation rate in the country with a high level of corruption. Subsidies in the agriculture sector further skew the value of forest conversion (e.g., see [50,51]). Finally, the soaring global demand for soy and beef within the context of tenure insecurity, high price of land and poor government regulation and enforcement, provide strong disincentives for forest restoration. However, increasing awareness combined with the involvement of financial entities in the sustainable development of the country, are putting pressure on producers to comply with the legislation in order to obtain their environmental licenses and be able to continue producing. For example, the government has prioritized the replication of the restoration projects that contribute to the legal mechanism "Conformance with the Forest Law" (CFL) currently being undertaken in four watersheds by the following public entities: SEAM, INFONA and the attorney general, with support from WWF.

In Madagascar on the other hand, a strong motivation by communities to engage in forest restoration is specifically the opportunity for enhanced tenure security (as emerged notably from our interviews). Through forest restoration (as well as improvements in forest management more generally), communities are empowered and recognized as rightful guardians of forests. Contracts established via the GELOSE or GCF mechanisms also help to secure titling and to exclude outsiders from the forest [57], thus reducing the risk of degradation and deforestation. Furthermore, in most cases, forest restoration in Madagascar is promoted via some form of payments (payment for ecosystem services, free seedlings through international projects *etc.*), providing an added incentive for communities to engage in forest restoration. For example, the Andasibe-Mantadia corridor project (see Table 3) provides payments to communities for both protecting and restoring the forest [60]. A final incentive in Madagascar that supports forest restoration is the reliance of communities on the provision of direct goods and services by the forest.

Regarding incentives, it is clear that, in both Madagascar and Paraguay, economics play a major role with respect to practicing restoration or not. In Paraguay, like in other parts of South America, restoration efforts must compete with lucrative agricultural commodities which are themselves promoted by subsidies [61], while in Madagascar, the household economy and the need to practice at least some subsistence agriculture impacts on decisions with respect to forest restoration. In both cases, economic incentives for restoration likely need to be increased in order to offer an alternative to the prevailing context.

Specific incentives that have influenced forest owners in both countries can be divided as legal, market, and financial. For example in Paraguay, legal incentives such as the requirement to set aside 25% of forest on private lands, combined with financial incentives, have encouraged private landowners to restore forests. At the same time, the high price of commodities such as soy has acted as a market disincentive in Paraguay, leading to forest conversion. In Madagascar, the provision of free seedlings to communities for them to engage in restoration in the Fandriana-Marolambo landscape has acted as an important incentive for them to not only restore the landscape but also to use a wider range of indigenous species. Payments for the ecosystem services of biodiversity conservation and carbon sequestration have been trialed in Madagascar (see examples in Table 3). Indeed, payments for ecosystem services, notably carbon sequestration and watershed protection, appear to be means that could enhance the appeal of forest restoration. For example, the REDD+ mechanism that is increasingly gaining ground provides payments for protection, effective management and restoration of forests (e.g., [2]).

3.3.2. Stakeholders

In addition to the processes dimension, the capacity, engagement, accountability and recognition of stakeholder groups also have a significant impact on governance of restoration zones and activities. The interplay at different levels, from local stakeholders to international actors, plays an important role in the governance of forests [56,64,65] and this can be seen with respect to forest restoration in both countries. For instance, in Paraguay, demand for soy from international players has a direct impact on local farmers' decisions to grow soy (at the cost of forest restoration). Equally in Madagascar, local communities' desire to recover their rights to manage forests is impacted by rules defined by national (and in some cases, such as the REDD+ mechanism, international) level players [65]. Furthermore, the role of "neutral" parties (such as international NGOs) appears critical in implementing management transfers in Madagascar (e.g., [66]). At the local level, tensions between local stakeholders, in particular communities or indigenous groups, *versus* private landowners, are particularly relevant in Paraguay (e.g., [13,61]) but also in Madagascar (e.g., [38,67]).

In both countries, the relative absence of the government forest service and the associated lack of enforcement of forest regulations have contributed to deforestation. In Madagascar, this has been counteracted to a certain extent by the empowerment of COBA associations which now have control of some forest areas, effectively curbing open access to these resources [44,67]. It appears that State authority and power have been largely absent in many areas in Paraguay's Atlantic Forest and Madagascar's forests and shrublands. Strengthening the presence of this key stakeholder group will likely lead to a more stable forest co-management regime in Madagascar and heightened respect for forest regulations and laws in Paraguay, ultimately enhancing restoration efforts.

We further argue that the lack of integration across levels (e.g., between local level structures and stakeholders and national processes and structures) as well as across the three dimensions of the framework (*i.e.*, between processes, stakeholders and structures) impedes the creation of an adequate governance context that is conducive to successful forest restoration. For example in Paraguay, complex forest legislation is not matched by supportive institutions at both the national and local levels to implement and enforce such legislation. In Madagascar on the other hand, all too frequently the reason for local-level engagement in co-management arrangements around forests (with or without a restoration dimension) is triggered by poor national level policies related to management and ownership rights. Resolution therefore of all three dimensions of the governance framework and particularly focusing at the national and local levels, would help provide a more positive climate for effective forest restoration.

Table 3. Projects involving restoration/reforestation for different objectives [a].

	Project	Aim/Type [b]	Ownership	Management
Madagascar	Makira—Makira Carbon Company (MCC) and Wildlife Conservation Society (WCS)	REDD [b]	Public	Management of Protected Area by International NGO; co-managed outside of the protected area
	Ankeniheny to Zahamena Forest Corridor (CAZ)—Conservation International	REDD	Public	Co-managed
	Fandriana to Vondrozo Forest Corridor (COFAV) Conservation International	REDD	Public	Co-managed
	Holistic Forest Conservation Programme (PHCF)—WWF and Good Planet	REDD	Public	Co-managed and Community management
	FORECA—GTZ/Inter-cooperation	REDD	Public	REDD+ readiness initiative (6 of the 8 project sites were in the forest and shrublands ecoregion)
	Andasibe-Mantadia Biodiversity Corridor—Conservation International	Carbon sequestration	Public	Co-managed and Community management
	WWF Fandriana Marolambo landscape	Ecological restoration	Public	Community management
Paraguay	TFCA (Tropical Forest Conservation Act) with Guyra Paraguay: Restoration of connectivity between two forest blocks in Caazapa National Park	Ecological restoration	Private	Debt swap board (includes NGOs and government)
	Restoration of four watersheds, developed by public entities: SEAM, INFONA and attorney general, with support from WWF; legal mechanism called "Conformance with the Forest Law" (CFL)	Ecological restoration	Private	Public
	ITAIPU bi-national: project called "Paraguay Biodiversity" with SEAM, the World Bank and Partners: to restore the connectivity of six protected areas and the ITAIPU Reserves along to the Parana River	Ecological restoration	Private	Public with support of NGOs
	Project called "Sustainable Management of Natural Resources." Ministry of Agriculture (KFW/GTZ). To develop an Agroforestry programme with small producers in the Oriental Region of Paraguay.	Agroforestry using mixed species	Community	Public
	Project of Conformance with the Forest Law. Coordinated by WWF, to restore 25% of forest reserve.	Ecological restoration	Private	Private and Public
	Project: Restoring the Monday Watershed. Project developed by Association "A Todo Pulmon Paraguay Respira".	Ecological and Agroforestry	Private	Private
	Restoration undertaken as a fine for those deforesting on their own land	Ecological restoration	Private	Public
	Restoration of areas to comply with sugar certification scheme	Ecological Restoration	Private	Private

Notes: Sources: [38,42,62,63]. [a] All in the two ecoregions being studied; Note that this list is not exhaustive but rather illustrative; [b] Reducing Emissions from Deforestation and Forest Degradation.

Table 4. Restoration/reforestation aims, practices and challenges for governance in Paraguay's Privately-Owned Forests and Madagascar's Co-managed State Forests

	Aim/Driver			Practices/Species	Challenges
	Ecological	Financial	Social/Political		
Madagascar	Carbon sequestration; Reducing forest degradation; Protection; Building resilience; Restoring forests; Protecting water	Subsidies and/or revenue; Revenue from carbon sequestration; Benefits from sustainable livelihood activities	Tenure security; Transfer of management rights; Employment; Provision of non-timber forest products (NTFPs)	**Restoration:** Native species; Useful tree species (e.g., for NTFPs); **Reforestation:** Fast growing exotic species for commercial purposes	**Governance challenges** Participation of communities; Legal transfer of management rights to communities; Lack of national policies to encourage restoration; Working with individual landowners; Tenure conflicts; Education and sensitization of populations; Government attitude to restoration **Other challenges** Cost; Managing native species effectively Technical difficulties; Communications to raise awareness among the population
Paraguay	Water protection; Reducing deforestation and forest degradation; Restoring forests	Marketable species; Revenue from carbon sequestration	Compliance with laws; Transfer of management rights	Rapid growth species; Species for which knowledge exists; Native species	**Governance challenges** Lack of proper control; Risk of illegal takeovers of land (tenure insecurity and property rights); Illegal logging; Absence of state; Lack of incentives; Poor political framework and support **Other challenges** Availability of seed producing trees; Lack of funds for maintenance

4. Conclusions

In conclusion, from a biodiversity perspective, the need for forest restoration in Paraguay's Atlantic Forests and Madagascar's forests and shrublands has been well established but success in this respect appears to be limited. One important factor contributing to this limitation is ongoing challenges related to governance. A three-pronged framework helped us to better understand the key issues and dimensions. Using this framework, it appears that the "processes" dimension of governance, which includes laws, strategies and incentives is particularly challenging, with poor policies and low implementation of the legislation. The "stakeholders" dimension, and in particular the interaction among stakeholders across different levels (from local to international), also appears to complicate effective governance of forest restoration in the two case studies.

The two case studies highlight that under different tenure arrangements, the governance challenges for forest restoration differed somewhat. So far, in Madagascar increased tenure security, provision of direct ecosystem goods and services, and payments for ecosystem services have been critical incentives for local community engagement in restoration. In contrast, in Paraguay, market pressures have provided a disincentive for forest restoration, and forest laws and regulations that favor restoration need to be applied with greater rigor. Nevertheless, in both cases weak government enforcement and remaining lack of clarity in tenure arrangements impede progress on forest restoration.

The present challenge in both countries lies in improving the forest governance context so that processes are more effective and key stakeholder groups can increase their participation in restoration activities. The lack of positive incentives is one of the main reasons for limited restoration activities, with the high prices of commodities being a significant disincentive for forest restoration. In both countries, enhanced economic incentives, such as payments for ecosystem services, are needed to contribute to a forest governance context that favors restoration.

Acknowledgments

The authors would like to thank the interviewees for taking the time to respond to our questions, Anjara Andriamanalina for designing the map of Madagascar, Menchi Garay for the map of Paraguay, Violeta Carrillo for helping with the bibliography for Paraguay, and the three anonymous reviewers and Anne Sgard for their valuable comments on an earlier draft.

Author Contributions

Stephanie Mansourian led on the general framework for the paper. Lucy Aquino contributed specific information on Paraguay while Francisco Pereira led on the interviews in Paraguay. Tom Erdmann contributed specific information on Madagascar and led on the interviews in Madagascar.

Conflicts of Interest

The views expressed represent those of the authors and do not necessarily represent those of their organizations.

References

1. Lamb, D.; Erskine, P.D.; Parrotta, J.A. Restoration of degraded tropical forest landscapes. *Science* **2005**, *310*, 1628–1632.
2. Kapos, V.; Kurz, W.A.; Gardner T.; Ferreira, J.; Guariguata, M.; Pin Koh, L.; Mansourian, S.; Parrotta, J.A.; Sasaki, N.; Schmitt, C.B. Impacts of Forest and Land Management on Biodiversity and Carbon. In *Understanding Relationships between Biodiversity, Carbon, Forests and People: The Key to Achieving REDD+ Objectives. A global Assessment Report*; Parrotta, J., Wildburger, C., Mansourian, S., Eds.; IUFRO: Vienna, Austria, 2012; pp. 53–80.
3. Society for Ecological Restoration International Science & Policy Working Group. The SER International Primer on Ecological Restoration. Available online: www.ser.org (accessed on 16 January 2014).
4. Bullock, J.M.; Aronson, J.; Newton, A.C.; Pywell, R.F.; Rey-Benayas, J.M. Restoration of ecosystem services and biodiversity: Conflicts and opportunities. *Trends Ecol. Evolut.* **2011**, *26*, 541–549.
5. Agrawal, A. Forests, governance, and sustainability: Common property theory and its contributions. *Int. J. Commons* **2007**, *1*, 111–136.
6. Dudley, N. Restoring Quality in Existing Native Forest Landscapes. In *Forest Restoration in Landscapes: Beyond Planting Trees*; Mansourian, S., Vallauri, D., Dudley, N., Eds.; Springer: New York, NY, USA, 2005.
7. Strassburg, B.N.; Vira, B.; Mahanty, S.; Mansourian, S.; Martin, A.; Dawson, N.M.; Gross-Camp, N.; Latawiec, A.; Swainson, L. Social and Economic Considerations Relevant to REDD+. In *Understanding Relationships between Biodiversity, Carbon, Forests and People: The Key to Achieving REDD+ Objectives, a Global Assessment Report*; Parrotta, J., Wildburger, C., Mansourian, S., Eds.; IUFRO: Vienna, Austria, 2012.
8. Lemos, M.C.; Agrawal, A. Environmental Governance. *Annu. Rev. Environ. Resour.* **2006**, *31*, 297–325.
9. Görg, C. Landscape governance the "politics of scale" and the "natural" conditions of places. *Geoforum* **2007**, *38*, 954–966.
10. United States Agency for International Development (USAID). *Guidelines for Assessing the Strengths and Weaknesses of Natural Resource Governance in Landscapes and Seascapes*; United States Agency for International Development: Washington, DC, USA, 2013. Available online: http://www.frameweb.org/CommunityBrowser.aspx?id=10650 (accessed on 16 January 2014).
11. Borrini-Feyerabend, G.; Dudley, N.; Jaeger, T.; Lassen, B.; Pathak Broome, N.; Phillips, A.; Sandwith, T. *Governance of Protected Areas: From Understanding to Action. Best Practice Protected Area Guidelines Series No. 20*; International Union for Conservation of Nature: Gland, Switzerland, 2013; p. 124.
12. Food and Agriculture Organization of the United Nations. *Forest Resources Assessment*; FAO: Rome, Italy, 2010.

13. Sunderlin, W.D.; Hatcher, J.; Liddle, M. *From Exclusion to Ownership? Challenges and Opportunities in Advancing Forest Tenure Reform*; Rights and Resources Initiative: Washington, DC, USA, 2008.

14. White, A.; Martin, A. *Who Owns the World's Forests*; Forest Trends and Center for International Environmental Law: Washington, DC, USA, 2002.

15. Agrawal, A.; Chhatre, A.; Hardin, R. Changing governance of the world's forests. *Science* **2008**, *320*, 1460–1462.

16. Kishor, N.; Rosenbaum, K. *Assessing and Monitoring Forest Governance: A User's Guide to a Diagnostic Tool*; Program on Forests (PROFOR): Washington, DC, USA, 2012. Available online: http://www.profor.info/node/1998 (accessed on 16 January 2014).

17. Davis, C.; Williams, L.; Lupberger, S.; Daviet, F. *Assessing Forest Governance*; World Resources Institute: Washington, DC, USA, 2013. (Available online at http://www.wri.org/publication/assessing-forest-governance (accessed on 16 January 2014).

18. Mansourian, S.; Oviedo, G. *Framework on Governance of Protected Areas and Livelihoods*; International Union for Conservation of Nature: Gland, Switzerland, 2009; p. 25.

19. Pahl-Wostl, C. A conceptual framework for analysing adaptive capacity and multi-level learning processes in resource governance regimes. *Glob. Environ. Chang.* **2009**, *19*, 354–365.

20. Myers, N.; Mittermeier, R.A.; Mittermeier, C.G.; da Fonseca, G.A.B; Kent, J. Biodiversity hotspots for conservation priorities. *Nature* **2000**, *403*, 853–858.

21. Olson, D.M.; Dinerstein, E. The Global 200: A representation approach to conserving the earth's most biologically valuable ecoregions. *Conserv. Biol.* **1998**, *12*, 502–515.

22. *Evolution de la Couverture de Forêts Naturelles à Madagascar 2005–2010*; Office National pour l'Environnement, Direction Générale des Forêts, Foiben-Taosarintanin'i Madagasikara, Madagascar National Parks, Conservation International: Antananarivo, Madagascar, 2013; p. 48.

23. United Nations Development Programme. *Human Development Report*; UNDP: New York, NY, USA, 2013.

24. Cartes, L.C. Breve historia de la conservación en el Bosque Atlántico. In *El Bosque Atlantico en Paraguay: Biodiversidad, Amenazas y Perspectivas*; Asociación Guyra Paraguay, Conservation International: Asuncion, Paraguay, 2005; pp. 37–54.

25. Aide, T.M.; Clark, M.L.; Grau, H.R; Lopez-Carr, D.; Levy, M.A.; Redo, D.; Bonilla-Moheno, M.; Riner, G.; Andrade-Nuñez, M.J.; Muñiz, M. Deforestation and Reforestation of Latin America and the Caribbean (2001–2010). *BIOTROPICA* **2013**, *45*, 262–271.

26. Hansen, M.C.; Potapov, P.V.; Moore, R.; Hancher, M.; Turubanova, S.A.; Tyukavina, A.; Thau, D.; Stehman, S.V.; Goetz, S.J.; Loveland, T.R.; *et al.* High-resolution global maps of 21st century forest cover change. *Science* **2013**, *342*, 850–853.

27. Roelens, J.B.; Vallauri, D.; Razafimahatratra, A.; Rambeloarisoa, G.; Razafy, F.L. *Restauration des paysages forestiers: Cinq ans de réalisation à Fandriana-Marolambo*; WWF France: Paris, France, 2010; p. 90.

28. Grinand, C.; Rakotomalala, F.; Gonde, V.; Vaudryc, R.; Bernoux, M.; Vieilledent, G. Estimating deforestation in tropical humid and dry forests in Madagascar from 2000 to 2010 using multi-date Landsat satellite images and the random forests classifier. *Remote Sens. Environ.* **2013**, *139*, 68–80.

29. Fogel, R.B. *La actual distribución de tierras en el Paraguay y el conflicto agrario*; *Documento Estudio N 4*; BaseInvestigaciones Sociales: Asunción, Paraguay, 1989.

30. Ribeiro, M.C.; Metzger, J.P.; Martensen, A.C.; Ponzoni, F.J.; Hirota, M.M. The Brazilian Atlantic forest: How much is left, and how is the remaining forest distributed? Implications for conservation. *Biol. Conserv.* **2009**, *142*, 1141–1153.

31. Di Bitetti, M.S.; Placci, G.; Dietz, L.A. *A Biodiversity Vision for the Upper Parana Atlantic Forest Ecoregion: Designing a Biodiversity Conservation Landscape Priorities for Conservation Action*; World Wildlife Fund: Washington, DC, USA, 2003; p. 104.

32. Paraguay Perfil Poblacional 2012. Available online: http://www.indexmundi.com/es/paraguay/poblacion_perfil.html (accessed on 16 January 2014).

33. Macedo, A.M.; Cartes, J.L. Aspectos Economicos del BAAPA. In *El Bosque Atlántico en Paraguay: Biodiversidad, Amenazas y Perspectivas*; Cartes, J.L., Ed.; Asociación Guyra Paraguay, Conservation International: Washington, DC, USA, 2005, pp. 107–126.

34. Guereña, A. *El Espejismo de la Soja: Los límites de la responsabilidad social empresarial*; Oxfam: Oxford, UK, 2013; p. 55.

35. White, F. *The Vegetation of Africa, a Descriptive Memoir to Accompany UNESCO/AETFAT Vegetation Map of Africa*; United Nations Educational, Scientific and Cultural Organization: Paris, France, 1983.

36. Goodman, S.M.; Benstead, J.P. *The Natural History of Madagascar*; Chicago University Press: Chicago, IL, USA, 2003.

37. Jarosz L. Defining and explaining tropical deforestation: Shifting cultivation and population growth in colonial Madagascar (1896–1940). *Econ. Geogr.* **1993**, *69*, 366–379.

38. Ferguson, B. REDD comes into fashion in Madagascar. *Madagascar Conserv.* **2009**, *4*, 132–137.

39. Bertrand, A. The Spread of the Merina People in Madagascar and Natural Forest and Eucalyptus Stand Dynamics. In *Beyond Tropical Deforestation*; Babin, D., Ed.; UNESCO/CIRAD: Paris, France, 2004.

40. Kull, C.A.; Ibrahim, C.K.; Meredith, T. Can Privatization Conserve the Global Biodiversity Commons? Tropical Reforestation Through Globalization. Presented at International Association for the Study of Common Property, Bali, 19–23 June 2006.

41. Davis, C.; Daviet, F.; Nakhooda, S.; Thuault, A. *A Review of 25 Readiness Plan Idea Notes from the World Bank Forest Carbon Partnership Facility*, WRI Working Paper; World Resources Institute: Washington, DC, USA, 2009. Available online: http://www.wri.org/gfi (accessed on 16 January 2014).

42. Wendland, K.J.; Honzák, M.; Portela, R.; Vitale, B.; Rubinoff, S.; Randrianarisoa, J. Targeting and implementing payments for ecosystem services: Opportunities for bundling biodiversity conservation with carbon and water services in Madagascar. *Ecol. Econ.* **2010**, *69*, 2093–2107.

43. Ferguson, B. Madagascar. In *REDD, Forest Governance and Rural Livelihoods: The Emerging Agenda*; Springate-Baginski, O., Wollenberg, E., Eds.; CIFOR: Bogor, Indonesia, 2010.

44. Hockley, N.J.; Andriamarovololona, M.A. The economics of community forest management in Madagascar: Is there a free lunch? An analysis of Transfert de Gestion. USAID: Washington, DC, USA, 2007.

45. Babin, D.; Bertrand, A. Comment Gérer le Pluralisme: Subsidiarité et Médiation Patrimoniale. *Unasylva* 1998. Available online: http://www.fao.org/DOCREP/W8827F/w8827f05.htm# comment gérer le pluralisme: subsidiarité et médiation part (accessed on 16 January 2014).

46. Anon, Ministère de l'Environnement et des Forêts, Antsiranana, Madagascar. Recommandations et Rapport d'Atelier, Journées Informatives et Prospectives sur les Transferts de Gestion des Ressources Naturelles. 2011, Unpublished work.

47. Randrianarisoa, A.; Raharinaivosoa, E.; Koll, H.E. In proceedings of Des Effets De La Gestion Forestière par les Communautés Locales De Base A Madagascar: Cas d' Arivonimamo et de Merikanjaka sur les Hautes Terres de Madagascar; Workshop on Forest Governance & Decentralization in Africa: Durban, South Africa, 8–11 April 2008.

48. Innes, J.L. Madagascar rosewood, illegal logging and the tropical timber trade. *Madagascar Conserv. Dev.* **2010**, *5*, 6–10.

49. Yanosky, A.; Cabrera, E. La capacidad Nacional de conservación del Bosque Atlántico. In *Biodiversidad, Amenazas y Perspectivas*; Cartes, J.L., Ed.; Asociación Guyra Paraguay, Conservation International: Asunción, Paraguay, 2005; pp. 137–172.

50. Contreras-Hermosilla, A. People, governance and forests—The stumbling blocks in forest governance reform in Latin America. *Forests* **2011**, *2*, 168–199.

51. McGinley, K.; Alvarado, R.; Cubbage, F.; Diaz, D.; Donoso, P.J.; Gonçalves Jacovine, L.A.; de Silva, F.L.; MacIntyre, C.; Monges Zalazar, E. Regulating the sustainability of forest management in the Americas: Cross-country comparisons of forest legislation. *Forests* **2012**, *3*, 467–505.

52. Vidal, V. Estudio sobre Mecanismos Financieros para el Manejo Forestal Sustentable en Sudamerica. Fase I. Cono Sur. FAO, 2004. Available online: http://www.rlc.fao.org/proyecto/ rla133ec/pag/i_paises.htm (accessed on 16 January 2014).

53. Chemonics International. *Tropical Forestry and Biodiversity Conservation in Paraguay: Final Report of a Section 118/119 Assessment EPIQ II Task Order No.1*; United States Agency for International Development: Asunción, Paraguay, 2004; p. 61.

54. Abed, S.R.; Santagada, E. *Régimen Jurídico Forestal de la República del Paraguay. Análisis crítico: Compilación normativa*; Instituto de Derecho y Economía Ambiental, Food and Agriculture Organization of the United Nations, Instituto Nacional Forestal: Asunción, Paraguay, 2011; p. 155.

55. WWF-Paraguay. *Ending Deforestation: Lessons Learnt from WWF's Experience*; WWF-Paraguay: Asuncion, Paraguay, 2007; p. 28.

56. Kanowski, P.J.; McDermott, C.L.; Cashore, B.W. Implementing REDD+: Lessons from analysis of forest governance. *Environ. Sci. Policy* **2011**, *14*, 11–117.

57. Jacoby, H.G.; Minten, B. Is land titling in sub-Saharan Africa cost-effective? Evidence from Madagascar. *World Bank Econ. Rev.* **2007**, *21*, 461–485.

58. Huang, C.; Kim, S.; Song, K.; Townshend, J.R.G.; Davis, P.; Altstatt, A.; Rodas, O.; Yanosky, A.; Clay, R.; Tucker, C.J.; Musinsky, J. Assessment of Paraguay's forest cover change using Landsat observations. *Glob. Planet. Chang.* **2009**, *67*, 1–12.

59. Wright, S.J.; Sanchez-Azofeifa, G.A.; Portillo-Quintero, C.; Davies, D. Poverty and corruption compromise tropical forest reserves. *Ecol. Appl.* **2007**, *17*, 1259–1266.

60. Jindal, R.; Swallow, B.; Kerr, J. Forestry-based carbon sequestration projects in Africa: Potential benefits and challenges. *Nat. Resour. Forum* **2008**, *32*, 116–130.

61. Pacheco, P.; Aguilar-Støen, M.; Börner, J.; Etter, A.; Putzel, L.; del Carmen Vera Diaz, M. Landscape transformation in tropical Latin America: Assessing trends and policy implications for REDD+. *Forests* **2011**, *2*, 1–29.

62. Borsy, P.; Sosa, E.V.; Molinas, W.; Cuellar, C.L.; Enciso, P. *Manejemos Nuestro Bosque: Manual de Manejo de Bosque Nativo en Pequeñas Fincas, Proyecto Manejo Sostenible de Recursos Naturales*; MAG/KfW/GTZ: Asunción, Paraguay, 2010; p. 46.

63. Hagen, R. *Evaluation des Projets Pilotes d'Aménagement des Forêts Naturelles A Madagascar*; United States Agency for International Development: Antananarivo, Madagascar, 2001.

64. Kothari, A.; Camill, P.; Brown, J. Conservation as if people mattered: Policy and practice of community-based conservation. *Conserv. Soc.* **2013**, *11*, 1–15.

65. Bidaud, C. REDD+, Un Mécanisme Novateur? Le cas de la forêt de Makira à Madagascar. *Rev. Tiers Monde* **2012**, *211*, 111–130.

66. McConnell, W.J.; Sweeney, S.P. Challenges of forest governance in Madagascar. *Geogr. J.* **2005**, *171*, 223–238.

67. Casse, T. The international debate on forest management transfer and our contribution. *Les Cah. d'Outre-Mer* **2012**, *257*, 11–46.

Appendix

Appendix 1. Questions posed to forest owners, managers and experts engaged in forest restoration in Paraguay and Madagascar.

1. Can you point to specific factors influencing your (others') decisions to restore or not forests in your country?

2. What is your/the primary motivation to restore forests?

3. What are the most common species used for restoration? What determines the choice of species for restoration/reforestation?

4. What determines the area chosen for restoration/reforestation?

5. What approaches/species are commonly used for restoration?

6. What could encourage you to restore more?

7. What are the challenges faced with forest restoration?

8. What are opportunities for forest restoration?

9. How successful would you rate forest restoration (in your area/country)? And on what are you basing your judgment?

10. How is the restored area currently managed/governed and how will it be managed/governed in the future? Who are the main actors in forest restoration management/governance and what is the relationship between these actors? What are the key challenges or opportunities with respect to governance of these areas?

Redefining Secondary Forests in the Mexican Forest Code: Implications for Management, Restoration, and Conservation

Francisco J. Román-Dañobeytia, Samuel I. Levy-Tacher, Pedro Macario-Mendoza and José Zúñiga-Morales

Abstract: The Mexican Forest Code establishes structural reference values to differentiate between secondary and old-growth forests and requires a management plan when secondary forests become old-growth and potentially harvestable forests. The implications of this regulation for forest management, restoration, and conservation were assessed in the context of the Calakmul Biosphere Reserve, which is located in the Yucatan Peninsula. The basal area and stem density thresholds currently used by the legislation to differentiate old-growth from secondary forests are 4 m^2/ha and 15 trees/ha (trees with a diameter at breast height of >25 cm); however, our research indicates that these values should be increased to 20 m^2/ha and 100 trees/ha, respectively. Given that a management plan is required when secondary forests become old-growth forests, many landowners avoid forest-stand development by engaging slash-and-burn agriculture or cattle grazing. We present evidence that deforestation and land degradation may prevent the natural regeneration of late-successional tree species of high ecological and economic importance. Moreover, we discuss the results of this study in the light of an ongoing debate in the Yucatan Peninsula between policy makers, non-governmental organizations (NGOs), landowners and researchers, regarding the modification of this regulation to redefine the concept of acahual (secondary forest) and to facilitate forest management and restoration with valuable timber tree species.

Reprinted from *Forests*. Cite as: Román-Dañobeytia, F.J.; Levy-Tacher, S.I.; Macario-Mendoza, P.; Zúñiga-Morales, J. Redefining Secondary Forests in the Mexican Forest Code: Implications for Management, Restoration, and Conservation. *Forests* **2014**, *5*, 978-991.

1. Introduction

Forest governance can be described as the *modus operandi* by which officials and institutions acquire and exercise authority in the management of forest resources. Good forest governance is characterized by predictable, open, and informed policymaking based on transparent processes; a bureaucracy imbued with a professional ethos; an executive arm of government that is accountable for its actions; and a strong civil society that participates in decisions related to the sector [1,2].

In 2003, the Mexican federal government published the Law for Sustainable Forestry Development (LSFD) with the primary objective of regulating and promoting the management, restoration, and conservation of forest ecosystems in the whole country [3]. This law authorizes timber harvesting in old-growth forest lands, and the establishment of commercial timber plantations in deforested lands. In 2005, the government published the regulation of this law [4]. The regulation determines the harvest potential based on specific minimum biomass/structural reference values that reflect the maturity of forest stands.

The definition of *acahual* (term used in Mexico for secondary forest) is only mentioned in the second point of the regulation and considers the native secondary vegetation that grows spontaneously in tropical forest lands that have previously been used for agriculture or cattle grazing. At this point, the regulation states that: (a) in evergreen or semi-evergreen forests, secondary vegetation is considered as those stands with less than 15 trees per hectare with a diameter at breast height (dbh) greater than 25 cm, or with a basal area less than 4 m^2/ha; and (b) in dry forests, secondary vegetation is considered as those stands with less than 15 trees per hectare with dbh greater than 10 cm, or with a basal area less than 2 m^2/ha [4].

In Mexico, the dominant vegetation communities are temperate forests, mostly *Pinus* and *Quercus* associations [5]. Therefore, it is possible that the LSFD and its regulation were developed based on the characteristics of these ecosystems, which are also the target of most of the commercial forestry in the country (65.3%) [6]. However, in the view of scientists and stakeholders from the Mexican tropical areas, this legislation is not sufficiently flexible to allow regional variations in best practice that would encourage innovation and experimentation. This is the case in the Yucatan Peninsula, an important source of tropical forest and non-forest products for the rest of the country. The authorities, ecologists, and landowners of this region have initiated a dialogue to review the implications of the LSFD for forest management, restoration, and conservation.

The Yucatan Peninsula encompasses the largest expanse of seasonal deciduous semi-evergreen tropical forest in Mexico, forming a complex and biodiversity-rich environmental gradient between the drier north of the peninsula and the humid Peten region in Guatemala [7]. Forest surveys performed in the Yucatan Peninsula have demonstrated the importance of the traditional slash-and-burn agriculture for forest regeneration, the recovery of soil fertility and the supply of secondary forest products (e.g., wood, construction materials, textiles, food, medicines, and tanning) that are vital for its rural economy [8,9]. In addition, tourism development in the Mexican Caribbean has increased the demand for palm leaves and round wood from secondary forests (<25 cm diameter at breast height, dbh), which are widely used for the construction of lodges and play a key role on the marketing of this tourism destination [10].

The basis for traditional secondary forest use and management in the Yucatan peninsula is tropical swidden agriculture (variously called shifting cultivation, slash-and-burn agriculture, or, in Mesoamerica, the milpa). Like most other tropical swidden systems, that of the peninsular Mayans centers on felling primary or secondary forest, burning the dried cuttings, and planting selected species in the clearing. Mayans plant and harvest a milpa for two to five consecutive years, then plant the area in tree crops and extracts fruit, rubber, and cordage as the fallowed area regenerates into secondary forest. When regrowth reaches a height of four to seven meters (usually within five to seven years), they clear and burn the area for a second round of cultivation, or allow it to regenerate into secondary forest, a process which requires approximately twenty years of fallowing [8,9].

In this study, we evaluated the accuracy of the biomass/structural reference values of the LSFD and its regulation for differentiating secondary from old-growth forests, and assessed whether they may be preventing the traditional use, management, and restoration of secondary forests and threatening the conservation of biodiversity in the Yucatan Peninsula. We also present our results in the light of an ongoing debate between the authorities, scientists, and practitioners of the Yucatan Peninsula to assess the implications of the current legislation on forest management, restoration, and conservation.

2. Materials and Methods

2.1. Study Area

The implications of the LSFD for forest management and conservation were assessed in the context of the Calakmul Biosphere Reserve (CBR), which is located in the state of Campeche, Yucatan Peninsula, Mexico (Figure 1). The CBR covers an area of 723,185 ha and is the largest tropical reserve in the country. Its topography is flat and smooth and its altitude varies from 260 to 385 m above sea level [11]. The climate is warm subtropical with a mean annual temperature of 24.6 °C and a mean annual rainfall of 1076 mm. Soils are shallow, calcareous, and highly permeable because of a high organic matter content and an underlying limestone bedrock [9].

Semi-evergreen forests cover most of the surface of the reserve. These are forests with trees reaching 15–25 m in height, 25%–50% of which lose their leaves during the dry season. The flora of Calakmul includes ~390 genera and 1500 species, 10% of which are endemic. The representative tree species of this type of forests are guayacán (*Guaiacum sanctum*), jobillo (*Astronium graveolens*), chicle (*Manilkara zapota*), ramón (*Brosimum alicastrum*), chakah (*Bursera simaruba*), and guarumo (*Cecropia obtusifolia*) [12].

Land tenure in the reserve is 49.6% communal, 48.4% property of the nation, and 2% privately owned [13]. The people who live within the reserve came from the states of Tabasco, Veracruz, Chiapas, and Michoacan, and their main activities involve slash-and-burn agriculture, cattle grazing and the harvesting of secondary forests [7]. These activities are complementary within the traditional (indigenous) system of shifting cultivation, in which managing forest fallows and second-growth forests is considered as a component of an integral agricultural system that relies on forest resilience [8,9].

2.2. Workshops

Between 2011 and 2013, representatives of the CBR funded and convened a total of 12 workshops to promote a multi-sectorial dialogue aimed at evaluating the potential implications of the LSFD and its regulation for traditional secondary forest management, as well as for the conservation of the region's biodiversity.

The participants of the workshops included representatives of the three broad groups that have a stake in ensuring good governance in the forest sector: (a) government: sub-national and national representatives of forest agencies and other departments and ministries; (b) civil society: representatives of community groups and social and environmental non-governmental organizations; and (c) the academic sector, represented by research specialists in forest ecology and management.

Representatives of the CBR encouraged scientists from El Colegio de la Frontera Sur to evaluate the accuracy of the forest ecological criteria stated in the regulation of LSFD. For this purpose, we conducted: (1) a review of the forest successional studies performed in the study region and in other similar tropical forests; (2) a comprehensive field sampling on forest successional development in the study area; and (3) an evaluation of the potential risks for biodiversity conservation that could stem from the implementation of the LSFD and its regulation.

Figure 1. Map of the study region in Southeastern Mexico, showing the location of the 50 forest plots sampled.

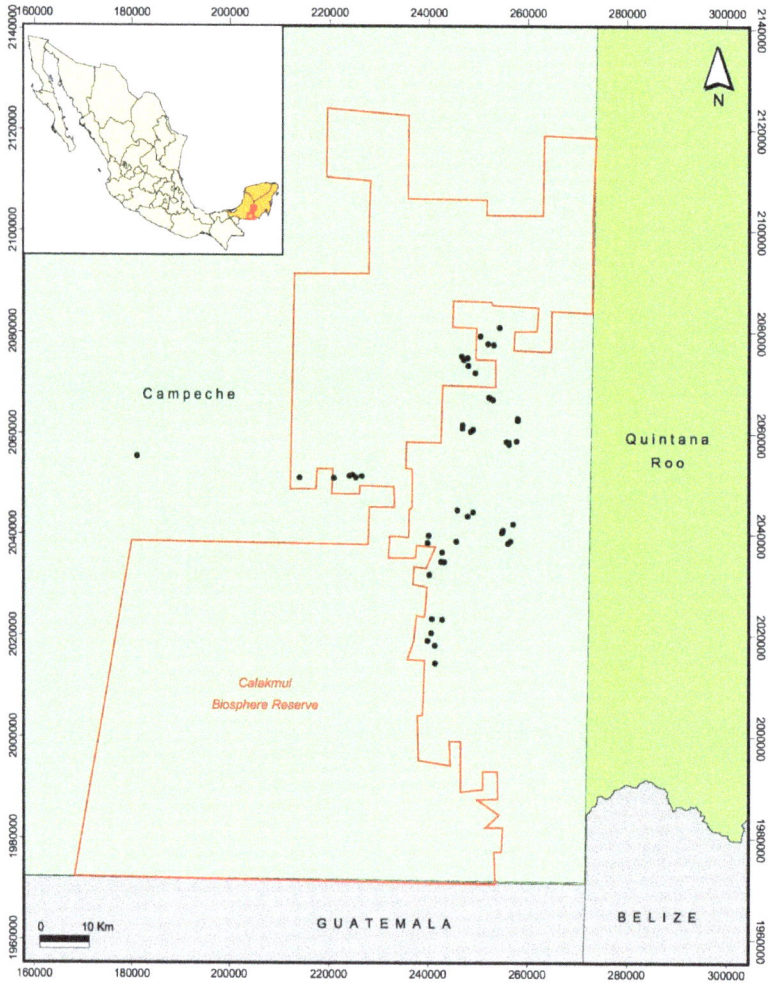

2.3. Field Sampling

To assess the reliability of the biomass/structural reference values established by the regulation of the LSFD for differentiating old-growth from secondary forests in the context of the semi-evergreen tropical forests of the Yucatan Peninsula, we performed vegetation assessments in the CBR using a chronosequence approach. Field data were recorded during plant surveys conducted during 2012. The surveys were based on a stratified random sampling design with a total of 50 sampling plots in five stages of forest succession (10 plots per age class), *i.e.*, 3–6, 9–11, 14–16, 19–21, and more than 50 years with no evidence of recent clearing, burning, or extractive human activities. The local authorities and informants from local communities helped to identify the tree species present in the plots, their main uses, and the land-use history of the different sites sampled.

After the identification of forest stands at different fallow intervals, we sampled 10 plots of 500 m^2 per age class. Using calipers, we measured all stems that were >2 cm dbh. The source of the species regeneration (seed or regrowth) was also recorded in a field notebook. Samples of the specimens were collected and deposited in the herbarium of El Colegio de la Frontera Sur, Chetumal headquarters. For species identification, we used dichotomous botanical keys, existing floristic lists for the study area [14], and sample contrast with herbarium specimens.

2.4. Data Analysis

Differences in basal area and stem density as a function of age were tested, via one-way analysis of variance (ANOVA). Tukey's multiple comparison procedure was applied if statistical differences were detected ($p < 0.05$). To comply with normality assumptions prior to ANOVA, stem density was $\log_{10}$ transformed [15]. Depending on species basal area across forest age classes, we classified species into successional groups, such as pioneer, persistent non-dominant, persistent dominant, and late-successional species [16]. We performed all statistical analyses and plots using the IBM SPSS Statistics processor, version 21.0.

3. Results and Discussion

3.1. Workshop Assessment

Workshops brought together 67 stakeholders from 10 organizations (including the government, civil society, and academia) with the aim of identifying possible inconsistencies and deliverable actions to improve the LSFD and its regulation. In general, the most recurrent problems in the legislation identified by the participants were: (1) the traditional use and management of secondary forests is not taken into consideration and has been relegated to illegality given that the reference values for distinguishing secondary from old-growth harvestable forests are controversial; (2) landowners prefer to dedicate resources to agriculture or cattle grazing than to forest management or restoration, to avoid complying with the costly management plans that are required by the legislation; and (3) the expansion of deforestation and land degradation may prevent the regeneration of slow-growing tree species of high economic value. As deliverable actions, participants agreed on the need to develop a reform proposal that should include: (1) the recognition of traditional secondary forest management for the provision of construction materials and other potential new forest products; (2) the modification of the reference values for distinguishing old-growth from secondary forests based on scientific data on the regional forest ecology; and (3) the development of a regional compensatory mechanism that supports forest restoration in harvested areas, especially regarding threatened species of high ecological and economic value.

3.2. Forest Successional Trends

In agreement with many studies on post agricultural tropical forest succession, our review of forest succession in the region indicated that basal area increases with time since abandonment, while stem density (stems >1 or 2 cm dbh) decreases with fallow age (Table 1).

In our study, basal area (F = 31.4, $p < 0.001$) and stem density (F = 67.0, $p < 0.001$) varied significantly among age classes. The minimum and maximum basal area values were 3.3 and 32.6 m^2/ha, respectively, and increased significantly with fallow age. These results reconfirmed that the reference value of 4 m^2/ha stipulated by the regulation of the LSFD was too low to distinguish old-growth from secondary forests appropriately, as young regenerating and secondary forest stands were all being considered as old-growth forests (Figure 2).

Stem density (>2 cm dbh) decreased significantly with age, from approximately 22,000 in the age class of 5 years to 4000 in the age class of 50 years (Figure 2). In contrast, stem density (>25 cm dbh) increased significantly with age, from zero in the youngest age class (5 years) to 218 in the age class of 50 years. These results indicate the presence of a large amount of thin stems in young stages and larger trees in more advanced successional stages. The reference value on stem density (15 trees > 25 cm dbh per hectare) of the regulation is also too low to reflect the structural differences between secondary and old-growth forests (Figure 2).

Table 1. Review of forest structural reference values among successional studies conducted in the Yucatan Peninsula and other similar tropical forests.

	Young Forest (<4–6 years)	Old-Growth Forest (30–50 years)	Forest Type	Location	Reference
Basal Area (m²/ha)	7.6	38.0	Semi-evergreen	Southern Yucatan	[9]
	7.5	17.0	Semi-evergreen	Eastern Yucatan	[17]
	7.6	22.7	Semi-evergreen	Southern Yucatan	This Study
	10.0	31.2	Seasonally dry	Central Yucatan	[18]
	5.0	22.5	Seasonally dry	Oaxaca	[19]
	15.0	28.0	Seasonally dry	Bolivia	[20]
Stem Density (#/ha)	20,000 (>1 cm dbh)	10,000 (>1 cm dbh)	Semi-evergreen	Southern Yucatan	[9]
	22,000 (>2 cm dbh)	4000 (>2 cm dbh)	Semi-evergreen	Southern Yucatan	This Study
	6638 (>2 cm dbh)	6644 (>2 cm dbh)	Seasonally dry	Central Yucatan	[18]
	5000 (>1 cm dbh)	4500 (>1 cm dbh)	Seasonally dry	Oaxaca	[19]
	8000 (>2 m tall)	4000 (>2 m tall)	Seasonally dry	Bolivia	[20]

3.3. Potential Risks for Biodiversity Conservation

Our field sampling also revealed that certain species might be threatened by the implementation of the LSFD and its regulation. The fact that landowners prefer to engage in agriculture and cattle grazing instead of forest management (as mentioned during the workshops) will lead to an increase in deforestation and land degradation. In this scenario, the natural regeneration of late-successional native tree species could be seriously threatened, given that, in our study, these species were absent in the early stages of succession (first 5–10 years); coexistence of late-successional species with pioneer and persistent species was only evident in more advanced successional stages (15–50 years) (Figure 3). Late-successional species are usually long-lived and shade-tolerant and produce large fruits and seeds that are dispersed mainly by mammals [21,22]. The representative late-successional species found in our sampling included *Brosimum alicastrum*, *Pimenta dioica*, *Talisia olivaeformis*, and *Manilkara sapota*.

Figure 2. Basal area and stem density in forest stands of Southern Yucatan, with abandonment times ranging from 3 to >50 years. The red dashed lines represent the current reference values established by the regulation of the Law for Sustainable Forestry Development (LSFD), whereas the blue dashed lines represent accurate science-based reference values. The different letters placed above the error bars indicate statistically significant differences (analysis of variance (ANOVA), Tukey's test, $p < 0.05$).

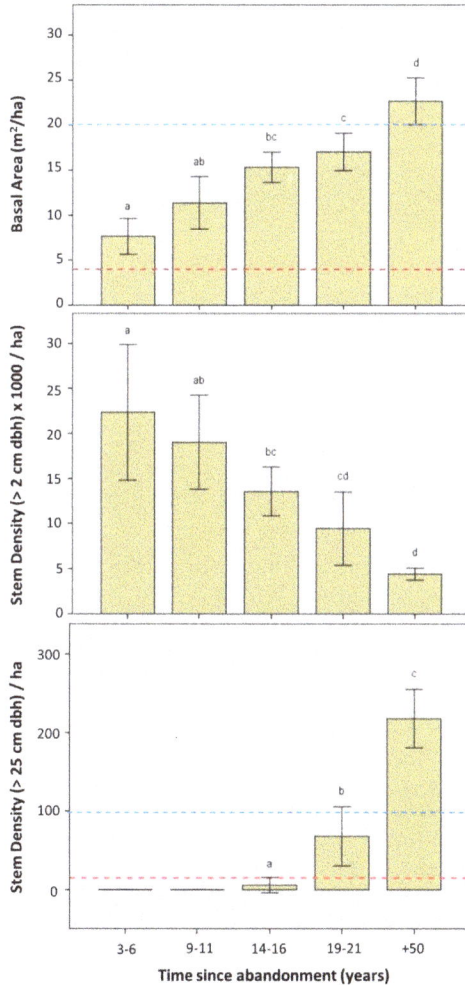

Figure 3. Basal area within successional groups across seasonal semi-evergreen forest stands, with abandonment times ranging from 5 to 50 years in the Yucatan Peninsula region.

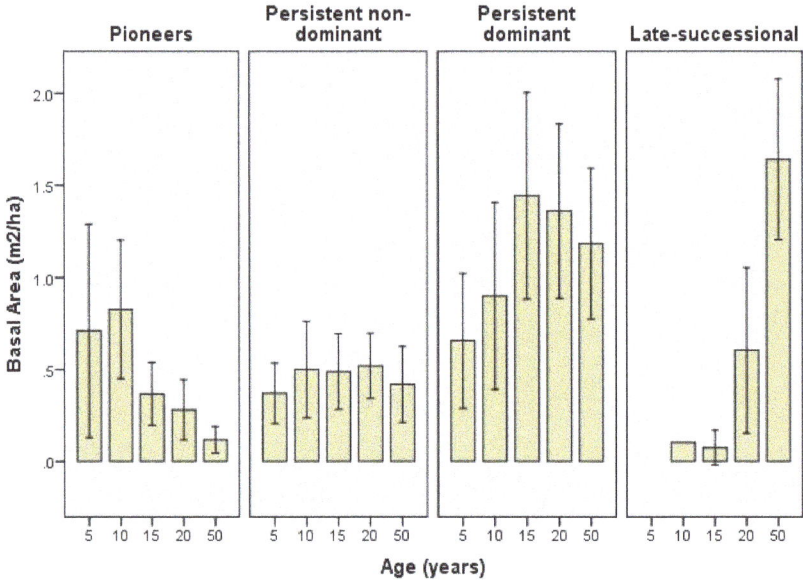

In contrast, pioneer species were dominant in the early stages of succession (first 5–10 years) and their abundance decreased progressively with time (Figure 3). Pioneer species are those that are able to colonize open areas; they are short-lived, fast-growing, and shade-intolerant species that produce small seeds that are dispersed mostly by the wind and birds [21,22]. The pioneer species found in our study included *Trema micrantha*, *Cecropia peltata*, *Solanum erianthum*, and *Hampea trilobata*.

In addition, persistent species were nearly constantly present throughout forest succession (Figure 3). The adaptation feature of persistent species to resprout from stumps (Table 2) enables them to survive disturbances (*i.e.*, slash-and-burn agriculture, hurricanes, and fire) and represent a high percentage of the initial floristic composition [23]. This allows the development of large-sized individuals over short periods, unlike what would happen if these species were established from seeds [9,24] as in the case of late-successional species (Table 2). In our study, persistent species were divided into dominant (e.g., *Bursera simaruba*, *Malmea depressa*, *Pouteria campechiana*, and *Dendropanax arboreus*) and non-dominant (e.g., *Guettarda combsii*, *Simarouba glauca*, *Piscidia piscipula*, *Coccoloba spicata*) species, depending on their basal area values across the successional stages.

Table 2. Number of species from different successional groups that regenerate from seeds and/or stump regrowth (data from the current study).

Successional Groups	Source of Regeneration	
	Stump Regrowth	Seeds
Pioneer Species	19	6
Persistent Non-dominant Species	37	5
Persistent Dominant Species	15	1
Late-Successional Species	3	14

3.4. Implications for Management, Restoration, and Conservation

One of the main problems detected by stakeholders regarding the LSFD and its regulation concerns the ignorance of the harvesting potential of secondary forests (trees < 25 cm dbh) for rural construction in Mayan communities and for the building of touristic rustic structures in the Mexican Caribbean. Also, non-timber forest products obtained from secondary forests such as the native palm species used for roofing (e.g. *Sabal yapa* and *Sabal mauritiiformis*), handicrafts (e.g., *Cryosophila argentea* and *Desmoncus quasillarius*), and ornamental (e.g. *Chamaedorea seifrizii*) have been a successful source of income for the rural communities of the Yucatan Peninsula, particularly as an alternative to high-value woods with larger diameters. However, the obligation to implement forest management plans (as those needed for large timber volumes) would render this activity non-viable economically. This requirement means that many landowners fail to have the necessary documentation for the marketing of these forest products. Nonetheless, all workshop participants agreed on the need of regulation of this activity to promote its legality and the conservation of Mayan traditional ecological knowledge.

Experiences worldwide have demonstrated that excessive regulations may entail prohibitively high transaction costs regarding legal operations, rendering adherence to the law impractical for many forest users [1,2]. This is particularly the case for community-based organizations from Southeast Mexico, which are often poorly equipped to comply with convoluted administrative procedures. Although there is strong evidence of a deep Mayan traditional ecological knowledge related to the sustainable use of secondary forests in the Yucatan Peninsula [25,26], this has been disregarded in the LSFD and its regulation, which have considered native forests as those that develop naturally without human intervention [3,4]. However, there is ample evidence of forest recovery after the slash-and-burn Mayan agricultural practice [8,9], which has shaped the Yucatecan forest landscape for centuries [23,27].

As a broad boundary between cultivated field and primary forest, the acahual contains a species structure and biomass distribution that differs from either field or forest, which facilitates the natural regeneration of many useful plants and late-successional tree species [25]. However, the acahual also plays an important role as a managed wildlife area since it contains a number of food sources not found in the forest that attract many animal species [28]. In fact, certain species seem to have adapted specifically to exploit this human-made niche, for they are found in larger numbers in acahual-bearing areas than they are in totally wild situations [7,10].

3.5. Future Perspectives

Our data on basal area and stem density were consistent with other studies that were conducted in similar forest types [29,30] and demonstrated that the reference values stipulated in the current Mexican forest code do not apply to the tropical forests of the Yucatan peninsula. Therefore, the values of forest structural criteria in the legislation should be increased, as suggested above. Forest reference values should be established based on reliable data and validated using participatory processes among users before publishing [2]. Otherwise, inconsistencies will lead to misinterpretations and failure to accomplish the goals of the law.

Our study also showed that the natural regeneration of late-successional species might be seriously threatened by the current problems detected in the LSFD and its regulation. Despite the expected rapid recovery of Yucatan forests because of their resprouting capability after clear-cutting activities [23], the establishment and further development of species that are key for long-term ecosystem functioning, such as long-lived and shade-tolerant late-successional tree species, are particularly vulnerable to deforestation and land degradation [22,26]. The seed germination and growth of late-successional species require specific sub canopy conditions of long fallow; thus these species are absent in young regenerating stands [31]. If not, their absence could severely affect the structure, composition, and functioning of forests [32]. In the long term, increased deforestation and the dominance of young regenerating stands at a landscape scale may lead to the regeneration of pioneer or persistent species exclusively, driving the loss of many late-successional species of the regional species pool [33].

Although the traditional shifting cultivation usually restarts the process of cutting down the forest again, it is necessary to manage at least a fraction of the secondary forests to become old-growth forests and to support the regeneration of threatened late-successional species of high ecological and social value. Legal instruments that encourage environmental certification for industries appear to be helpful for funding forest management and restoration activities in rural communities of tropical countries [2,34]. Valuation via the promotion of the trading and selling of certified forest products and rewards for ecosystem services rendered (including carbon sequestration) can also help increase forest conservation [35,36]. Similarly, the capacity of the Southern Yucatan forests to provide critical ecosystem services to thousands of people and to generate jobs for the inhabitants of local communities via forest management and restoration can be improved by specific adjustments of the current Mexican Forest Code.

4. Conclusions

The current forest structural reference values established by the regulation of the Mexican LSFD are controversial and do not apply to the tropical forests of the Yucatan, as they do not serve to differentiate between young regenerating, secondary, and old-growth forests appropriately. Moreover, the LSFD and its regulation disregard the traditional harvesting of secondary forests for rural construction, thus forcing small landowners to comply with management plans as if they were major timber producers. Although the stump regrowth of persistent species is important for forest regeneration after agricultural land use in the Yucatan peninsula, the diversity and coexistence of species from different successional groups is maximized in intermediate successional stages (about 20 years after abandonment). Late-successional species are particularly vulnerable to increased deforestation and the systematic clear cutting of young secondary forests, because of the specific conditions that are required for their regeneration, which can only be achieved through long fallow periods (>20 years). The facilitation of participatory processes between the different stakeholders involved in local forestry allowed the review of the major problems of the LSFD and its regulation, as well as the implementation of a field sampling to evaluate the accuracy and implications of the forest ecological criteria of these legal instruments. To adjust the existing reference values, as well as to develop a consensual concept of secondary forests, both in the regulation and in the law itself, modifications should be accompanied by accessible and well-documented procedures, and by fiscal incentives to encourage voluntary investments in forest management and restoration.

Acknowledgments

We thank the members of the Regional Association of State Foresters and Etnobiología para la Conservación for technical field support. We also thank the Calakmul Biosphere Reserve for the funds provided for this study. We greatly appreciate the useful comments of the reviewers and editors of the Special Issue on the earlier drafts of this manuscript.

Author Contributions

Francisco Román led the general organization of the paper. Samuel Levy contributed information on species ecology and successional groups. Pedro Macario contributed to traditional secondary forest management and led the field sampling. José Zúñiga contributed specific information on Calakmul and led the organization of the workshops. All co-authors contributed to the general framework of the paper.

Conflicts of Interest

The authors declare no conflict of interest.

References

1. Food and Agriculture Organization of the United Nations (FAO); International Tropical Timber Organization (ITTO). *Forest Governance and Climate-Change Mitigation*; Policy Brief, FAO/ITTO Initiative on Forest Law Compliance and Governance: Rome, Italy, 2009.
2. Aronson, J.; Brancalion, P.H.S.; Durigan, G.; Rodrigues, R.R.; Engel, V.L.; Tabarelli, M.; Torezan, J.M.D.; Gandolfi, S.; de Melo, A.C.G.; Kageyama, P.Y.; *et al.* What role should government regulation play in ecological restoration? Ongoing debate in Sao Paulo State, Brazil. *Restor. Ecol.* **2011**, *19*, 690–695.
3. Comisión Nacional Forestal (CONAFOR). *Ley General de Desarrollo Forestal Sustentable*; Diario Oficial de la Federación (25 Febrero): México D.F., Mexico, 2003. Available online: http://www.diputados.gob.mx/LeyesBiblio/pdf/259.pdf (accessed on 12 June 2013).

4. Comisión Nacional Forestal (CONAFOR). *Reglamento de la Ley General de Desarrollo Forestal Sustentable*; Diario Oficial de la Federación (21 Febrero): México D.F., Mexico, 2005. Available online: http://www.diputados.gob.mx/LeyesBiblio/regley/Reg_LGDFS.pdf (accessed on 12 June 2013).

5. Instituto Nacional de Estadística, Geografía e Informática (INEGI). *Información Geográfica*; INEGI: México D.F., Mexico, 2005.

6. Secretaría de Agricultura y Recursos Hidráulicos (SARH). *Inventario Nacional Forestal Periódico (1992–1994)*; SARH: México DF, Mexico, 1994.

7. Vester, H.F.M.; Lawrence, D.; Eastman, J.R.; Turner, B.L., II; Calme, S.; Dickson, R.; Pozo, C.; Sangermano, F. Land change in the southern Yucatan and Calakmul Biosphere Reserve: Effects on habitat and biodiversity. *Ecol. Appl.* **2007**, *17*, 989–1003.

8. Levy-Tacher, S.I.; Aguirre-Rivera, J.R. El aprovechamiento agrícola intensivo de los Hubchés (Acahuales o comunidades secundarias) de Yucatán. *Rev. Geogr.* **2000**, *28*, 79–103.

9. Read, L.; Lawrence, D. Recovery of biomass following shifting cultivation in dry tropical forests of the Yucatan. *Ecol. Appl.* **2003**, *13*, 85–97.

10. Ellis, E.A.; Porter-Bolland, L. Is community-based forest management more effective than protected areas? A comparison of land use/land cover change in two neighboring study areas of the Central Yucatan Peninsula, Mexico. *For. Ecol. Manag.* **2008**, *256*, 1971–1983.

11. Instituto Nacional de Estadística, Geografía e Informática (INEGI). *Campeche: Datos por Ejido y Comunidad Agraria. XI Censo General de Población y Vivienda, 1990*; INEGI: Aguascalientes, México DF, Mexico, 1996.

12. Martínez, E.; Galindo-Leal, C. La vegetación de Calakmul, Campeche, México: Clasificación, descripción y distribución. *Bol. Soc. Bot. Méx.* **2002**, *71*, 7–32.

13. Sánchez-González, M.C. Calakmul, encuesta. *Proyecto de Evaluación de Áreas Naturales Protegidas de México*; Gómez-Pompa, A., Dirzo, R., Eds.; Secretaría de Desarrollo Social, Sedesol: México DF, Mexico, 1993.

14. Pennington, T.D.; Sarukhán, J. *Árboles Tropicales de México*, 3rd ed.; Universidad Nacional Autónoma de México y Fondo de Cultura Económica: México DF, Mexico, 2005.

15. Fry, J.C. *Biological Data Analysis. A Practical Approach*; Oxford University Press: Oxford, UK, 1993.

16. Levy-Tacher, S.I.; Aguirre-Rivera, J.R. Successional pathways derived from different vegetation use patterns by Lacandon Mayan Indians. *J. Sustain. Agric.* **2005**, *26*, 49–82.

17. Urquiza-Haas, T.; Dolman, P.M.; Peres, C.A. Regional scale variation in forest structure and biomass in the Yucatan Peninsula, Mexico: Effects of forest disturbance. *For. Ecol. Manag.* **2007**, *247*, 80–90.

18. Hartter, J.; Lucas, C.; Gaughan, A.E.; Aranda, L.L. Detecting tropical dry forest succession in a shifting cultivation mosaic of the Yucatan Peninsula, Mexico. *Appl. Geogr.* **2008**, *28*, 134–149.

19. Lebrija-Trejos, E.; Pérez-García, E.A.; Meave, J.A.; Bongers, F.; Poorter, L. Functional traits and environmental filtering drive community assembly in a species-rich tropical landscape. *Ecology* **2010**, *91*, 386–398.

20. Kennard, D.K. Secondary forest succession in a tropical dry forest: Patterns of development across a 50-year chronosequence in lowland Bolivia. *J. Trop. Ecol.* **2002**, *18*, 53–66.

21. Whitmore, T.C. Canopy gaps and the two major groups of forest trees. *Ecology* **1989**, *70*, 536–538.

22. Gourlet-Fleury, S.; Blanc, L.; Picard, N.; Sist, P.; Dick, J.; Nasi, R.; Swaine, M.; Forni, E. Grouping species for predicting mixed tropical forest dynamics: Looking for a strategy. *Ann. For. Sci.* **2005**, *62*, 785–796.

23. Negreros-Castillo, P.; Hall, R.B. Sprouting capability of 17 tropical tree species after overstory removal in Quintana Roo, Mexico. *For. Ecol. Manag.* **2000**, *126*, 399–403.

24. Clarke, P.J.; Lawes, M.J.; Midgley, J.J.; Lamont, B.B.; Ojeda, F.; Burrows, G.E.; Enright, N.J.; Knox, K.J.E. Resprouting as a key functional trait: How buds, protection and resources drive persistence after fire. *New Phytol.* **2013**, *197*, 19–35.

25. Hernández-Xolocotzi, E.; Bello, E.; Levy-Tacher, S. *La Milpa en Yucatán, un Sistema de Producción Agrícola Tradicional*; Colegio de Postgraduados: Montecillos, México, 1995.

26. Hernandez-Stefanoni, J.L.; Dupuy, J.M. Effects of landscape patterns on species density and abundance of trees in a tropical subdeciduous forest of the Yucatan Peninsula. *For. Ecol. Manag.* **2008**, *255*, 3797–3805.

27. Zidar, C.; Elisens, W. Sacred giants: Depiction of Bombacoideae on Maya ceramics in Mexico, Guatemala, and Belize. *Econ. Bot.* **2009**, *63*, 119–129.

28. Nations, J.D.; Nigh, R.B. The evolutionary potential of Lacandon Maya sustained-yield tropical forest agriculture. *Anthr. Res.* **1980**, *36*, 1–30.

29. Guariguata, M.; Ostertag, R. Neotropical secondary forest succession: Changes in structural and functional characteristics. *For. Eco. Manag.* **2001**, *148*, 185–206.

30. Bonilla, M. Forest Recovery and Management Options in the Yucatan Peninsula, Mexico. Ph.D. Thesis, University of California, Santa Cruz, CA, USA, 2008.

31. Wright, S.J. Plant diversity in tropical forests: A review of mechanisms of species coexistence. *Oecologia* **2002**, *130*, 1–14.

32. Isbell, F.; Calcagno, V.; Hector, A.; Connolly, J.; Harpole, W.S.; Reich, P.B.; Scherer-Lorenzen, M.; Schmid, B.; Tilman, D.; van Ruijven, J.; *et al.* High plant diversity is needed to maintain ecosystem services. *Nature* **2011**, *477*, 199–203.

33. Martinez-Garza, C.; Howe, H.F. Restoring tropical diversity: Beating the time tax on species loss. *J. Appl. Ecol.* **2003**, *40*, 423–429.

34. Rodrigues, R.R.; Gandolfi, S.; Nave, A.G.; Aronson, J.; Barreto, T.E.; Vidal, C.Y.; Brancalion, P.H. Large-scale ecological restoration of high-diversity tropical forests in SE Brazil. *For. Ecol. Manag.* **2011**, *261*, 1605–1613.

35. Alexander, S.; Nelson, C.; Aronson, J.; Lamb, D.; Martinez, D.; Harris, J.; Higgs, E.; Lewis, R.R., III; Finlayson, M.; Erwin, K.; *et al.* Opportunities and challenges for ecological restoration within REDD+. *Restor. Ecol.* **2011**, *19*, 683–689.

36. Wu, T.; Kim, Y.-S.; Hurteau, M.D. Investing in natural capital: Using economic incentives to overcome barriers to forest restoration. *Restor. Ecol.* **2011**, *19*, 441–445.

Governing Forest Landscape Restoration: Cases from Indonesia

Cora van Oosten, Petrus Gunarso, Irene Koesoetjahjo and Freerk Wiersum

Abstract: Forest landscape restoration includes both the planning and implementation of measures to restore degraded forests within the perspective of the wider landscape. Governing forest landscape restoration requires fundamental considerations about the conceptualisation of forested landscapes and the types of restoration measures to be taken, and about who should be engaged in the governance process. A variety of governance approaches to forest landscape restoration exist, differing in both the nature of the object to be governed and the mode of governance. This paper analyses the nature and governance of restoration in three cases of forest landscape restoration in Indonesia. In each of these cases, both the original aim for restoration and the initiators of the process differ. The cases also differ in how deeply embedded they are in formal spatial planning mechanisms at the various political scales. Nonetheless, the cases show similar trends. All cases show a dynamic process of mobilising the landscape's stakeholders, plus a flexible process of crafting institutional space for conflict management, negotiation and decision making at the landscape level. As a result, the landscape focus changed over time from reserved forests to forested mosaic lands. The cases illustrate that the governance of forest landscape restoration should not be based on strict design criteria, but rather on a flexible governance approach that stimulates the creation of novel public-private institutional arrangements at the landscape level.

Reprinted from *Forests*. Cite as: van Oosten, C.; Gunarso, P.; Koesoetjahjo, I.; Wiersum, F. Governing Forest Landscape Restoration: Cases from Indonesia. *Forests* **2014**, *5*, 1143-1162.

1. Introduction

Forest landscape restoration (FLR) is rapidly gaining ground as an integrated approach towards allocating and managing land to achieve social, economic, and environmental objectives in areas where agriculture, mining, and other productive land uses compete with environmental and biodiversity goals [1]. Active lobbying by international organisations has led to FLR being integrated into international commitments such as the *Reducing Emissions from Deforestation and Forest Degradation* (REDD) arrangements identified by the UN Forum on Forests, the Aichi target No. 15 of the Convention on Biodiversity aiming to restore 15% of degraded ecosystems, and the Bonn Challenge, which aims to restore 150,000,000 ha by 2020 [2]. As part of the Bonn Challenge, an increasing number of governments have been pledging part of their national territory to be restored, and national assessments of the potential are currently being carried out looking at where and how these pledged areas could best be situated [3].

Although the FLR approach is formally recognised, many FLR programmes are still experimental in nature. In general terms, FLR refers to restoring the ecological services of forests within landscapes: not necessarily by bringing them back to their original state, but by restoring their functionality in terms of biodiversity, ecological functioning, livelihoods, or income [1].

Despite global efforts and ambitious targets for such attempts to reconcile conservation and development, there are as yet no general and effective solutions for meeting both nature conservation and human needs. The main reason is that the competing demands on land for conservation and development imply inevitable trade-offs, and there still is no unambiguous framework for how best to guide the process of decision making and implementation of forest restoration at the landscape level. Sometimes it is assumed that forest landscape restoration can be approached as a professional planning exercise, based on the idea that international and national targets "naturally" trickle down through the spatial planning systems of states. However, it is increasingly acknowledged that these politically and administratively oriented planning mechanisms do not always tally with the socio-ecological identity of forested landscapes. Several authors [2,4,5] have recognised the shortcomings of formal governance structures and their relative inability to govern restoration at the landscape level. These authors see the restoration process as involving "living" forest landscapes that are shaped by multiple social actors and networks, who operate across the bureaucratic sectorial and scaled planning structures of states. The landscape provides its inhabitants with the basis for their sociocultural and production practices, which in turn provide the institutional space for governance mechanisms to emerge. Consequently, forest landscape restoration involves multi-actor networks composed of people living in the landscape or indirectly belonging to it and requires new forms of planning and implementation of socio-ecological complexes. Such new forms of landscape governance should be characterised by (1) a geographical focus, integrating multiple sectors (agriculture, forests, water, *etc.*) within a single space; (2) a multi-actor focus bringing together public and private actors operating within a shared space; and (3) operating at multiple scales, meaning that they stretch across local, regional and global networks of spatial decision making, sometimes referred to as "politics of scale" [2,4,5]. Based on these principles, Sayer *et al.* [1] identified 10 major design principles for a landscape governance approach, including multi-functionality of landscapes, multi-level and multi stakeholder involvement, the importance of a shared concern, strengthened stakeholder capacity, negotiated and transparent change logic, clarification of rights and responsibilities, and continual learning and adaptive management. These principles are still rather generic, as they do not specify whether and how they are related to the two major critical issues in forest landscape restoration, *i.e.*, the object of governance and the nature of the governance process (*cf.* [6,7]). As a result of the multidimensional nature of the FLR governance process and the generic nature of the identified design principles, there is still a great deal of variation in the way FLR programmes are planned and implemented in practice. Consequently, further understanding is needed of the multiple interpretations of the concept of governing forest landscape restoration.

This article aims to contribute towards a better understanding of the nature and diversity of the process of forest restoration governance in terms of the object to be governed and the nature of the governance process. It takes the reader through an analytical framework based on (a) the different interpretations of forested landscapes and their relevant forms of restoration, and (b) the various modes of governance for steering decision making at the landscape level. Combining these two, the authors claim that the governance of forest landscape restoration can be regarded as a management tool; as a multi-stakeholder decision making process; or as the creation of new institutional space for spatial decision making. These three modes of governance are illustrated by three cases of forest landscape restoration in Indonesia, which are governed in different ways, depending on the gradual changes in both the substance and the modes of governance, which emerge out of their local realities.

2. Analytical Framework

Although the concept of forest landscape restoration is relatively new, the notion of the need to restore degraded and deforested landscapes is a long-standing one. As early as the mid-20th century, this notion resulted in programmes for watershed management and reforestation of degraded (or wasted) forest lands [8–10]. These "first-phase" forest restoration programmes were based on concerns about the loss of forest functions with respect to hydrological regulation, soil conservation and timber production. These programmes focused both on rehabilitation of denuded forest lands as well as erosion control and agroforestry development on the adjacent private agricultural lands. Gradually, the interpretation of forest degradation was extended to include a larger variety of forest services, such as supporting, regulating, provisioning and cultural services [11,12]. As a result, attention within forest restoration gradually shifted from the original emphasis on watershed services to a larger complex of ecological services, and understanding of the multiple manifestations of restored forests widened [13]. One repercussion of this development was that the concept of forest landscape restoration became more holistic and inclusive on the one hand, but it strengthened the forest focus on the other, with less attention being paid to adjacent agricultural lands. At the same time, the interpretation of the best approach to forest landscape governance and the related approaches to decision making and implementation also changed. Initially, an administrative and professional approach predominated, but gradually a multi-level and multi-actor governance approach evolved. Consequently, when considering the actual nature of forest restoration programmes and their governance, divergent interpretations can be identified in terms of (1) the substance of the governance process with respect to the type of forested landscapes and related forms of restoration; and (2) the modes of governance for steering decision-making at the landscape level.

2.1. Types of Forested Landscapes and their Relevant Form of Restoration

The notion of a "forested landscape" is open to various interpretations. On the one hand, it may be interpreted in an ecological sense as referring to a complex of different forest ecosystems which are integrated in a natural ecological structure, allowing good provision of ecological services and good distribution and dispersal of biodiversity. Alternatively, it may be interpreted from a socio-geographical perspective as referring to a spatial unit of land with a mosaic of forest and agricultural fields, created by local people as part of their livelihood activities. These mosaics often include a variety of forest types ranging from natural forests to various forms of anthropogenically modified forests, the latter also being referred to as rural or domestic forests [14–16]. These different interpretations of forested landscapes imply different approaches towards their restoration. The first interpretation leads to a restoration which focuses predominantly on restoring the ecological structure and environmental services of the forests as natural ecosystems. It is recognised here that ecological restoration improves the environmental services that forests provide for the various stakeholders, but little attention is paid to the question of how these services are delivered to the intended beneficiaries [13]. In contrast, the second interpretation leads to the recognition that forest landscape restoration often takes place in areas where forests have been adapted to human needs and where agriculture and other productive land uses compete with the environmental and biodiversity goals of restoring the forests. The second interpretation therefore considers not only how to ecologically restore forests, but also how to optimise the interactions between forests and other forms of land use. This offers scope for focusing not only on the restoration of natural forests, but also on anthropogenically modified forests and agrarian lands that are incorporated into forest mosaic landscapes.

This latter issue raises the question of what the role of people in the forest landscape is. Although forest degradation is the result of human exploitation of forests, it does not mean that local people should be considered as mere environmental degraders, who should be removed from the forest landscape; people can also act as an aggrading rather than degrading force in forested areas [17]. Such human agency is illustrated by the many creative examples of hybrid and sustainable human/nature systems in the form of rural (or domesticated) forests, managed by local people [18,19]. Such adapted forests, in which the provisioning services for local use have been optimised, indicate the potential for developing ecologically healthy landscapes with forests types that are adjusted to the needs of the inhabitants. Forest mosaic landscapes consisting of a mix of natural forests, adapted forests and agrarian land often provide better human living conditions than extended natural forest reserves, which implies that restoration of forested landscapes may imply more than the restoration of forests [20].

2.2. Modes of Governance for Steering Landscape Decision-Making

Forest landscape restoration concerns not only the implementation of a specific set of technical and ecological practices for developing a specific type of restored forests, but also the design, the planning and the decision-making at crucial moments during the process [1]. It is generally agreed that this process is quite complex, due to the nature of a landscape as involving multiple land uses and multiple stakeholders. In particular, the restoration of mosaic landscapes usually requires participation of the stakeholders involved in the various landscape components. The process even becomes more complex when landscapes stretch across political and administrative boundaries, and therefore cover more than one administrative planning unit. Whereas the initial watershed management projects mainly involved forestry agencies and local communities, in the current forest landscape restoration programmes, a much larger variety of stakeholders are recognised, including commercial enterprises. Moreover, the increased focus on a variety of forest services has resulted in increasing numbers of sectorial regulations and guidelines that need to be taken into consideration.

As a result, it is becoming increasingly recognised that landscape restoration requires the involvement of multiple stakeholders operating in multiple sectors, and at multiple scales. This type of stakeholder involvement in design, planning and decision-making of forest landscape restoration programmes is increasingly referred to by the term "landscape governance" [2,4]. During the last decade, the concept of landscape governance has become generally accepted as referring to the multi-stakeholder process of negotiation and decision making about policies and programmes for effective conservation and sustainable use of forests, and for implementing the planned measures within spatial landscape units [2,5,21]. Despite this general acceptance, there still is divergence in the way landscape governance is perceived and implemented in different restoration programmes. Treib *et al.* [22] identify different modes of governance with respect to the three different dimensions of politics, polity and policy. The modes of governance in the political dimension are related to whether only public actors are involved or also private ones (the actor constellation). The modes of governance in the polity dimension may vary, depending on whether they are based on a hierarchical government or a market approach; on a central locus of authority *versus* dispersed loci of authority; or on institutionalised *versus* non-institutionalised interactions (the institutional properties). The modes of governance in the policy dimension are related to whether the process is based on legally binding rules or on soft law; on a rigid approach to implementation *versus* a flexible one; on the presence or the absence of sanctions; and on material *versus* procedural regulation (the steering instruments). Deriving from these ideas, the authors conclude that three main modes of governance may be identified within forest landscape governance, *i.e.*, landscape governance primarily as a management tool; landscape governance as a multi-stakeholder decision making process; and landscape governance as the creation of new institutional space for spatial decision making.

Landscape governance as a management tool is still based on a rather traditional hierarchical system of decision making based on a central locus of authority, professional knowledge, binding regulations and a rather rigid approach to implementation. This does not mean in practice that stakeholder interaction may be less rigid, and management responsibilities may be shared. Such sharing of responsibilities is generally considered to be more effective than straightforward governmental control, as it increases a feeling of responsibility among landscape users and provides an opportunity to incorporate location-specific information. Sharing of responsibilities is also seen as an effective tool for mitigating conflicts, as it helps improve relationships between governments, private actors and a landscape's inhabitants. This interpretation of landscape governance is closely related to the concepts of co-management and collaborative management that are frequently applied in the local management of forest resources [4]. Stakeholders can be trained as co-managers in implementing management techniques, and made jointly responsible for the results. This is especially relevant to conservation agencies that plan forest restoration programmes on the forest lands they own.

Landscape governance as a process of multi-stakeholder decision-making is a mode of governance that pays attention specifically to the formation of new institutional interactions with increased scope for private actors and a flexible soft law approach to stimulating location-specific landscape practices rather than just implementing professional practices. This governance mode is often adopted in programmes covering complex mosaics of different land uses, where management involves a process of delicate and politically oriented decision making concerning preferred land use, paying attention not only to the rules, regulations and practices from the forest sector, but also to those from the agricultural sector. Multi-stakeholder decision making thus becomes a complex process of negotiation, conflict mediation and trade-offs [4,23]. This process is often conflictive in nature and needs careful facilitation and procedural management. Decisions about different land uses involve not only the direct stakeholders but also the complex networks they represent; networks that may transcend the boundaries of sectors and scales. There is a need here to recognise the different power positions of stakeholders operating from various sectors and scales, influenced by institutional drivers related to access to resources, as well as external drivers such as global market forces.

Landscape governance as the creation of institutional space is a mode of governance that allows more power for the private actors and market forces within the governance process. This requires more flexible forms of institutionalisation and implementation, especially in cases where landscapes are not restricted to a specific level in the spatial decision making structures of the state bureaucracy (provincial, district, or municipal level). Where landscapes stretch across administrative boundaries and political entities, multi-stakeholder decision making at the landscape level is hampered by the absence of spatial decision-making structures embedded in formal institutional frameworks. These cases illustrate the fact that landscapes are socio-ecological constructs, shaped and reshaped by landscape actors themselves, stretching beyond the planning structures of states. In such cases, landscape governance cannot be the outcome of formal planning structures, but is rather the outcome of "institutional bricolage": landscape actors from different sectors and scales create new institutional space by creatively combining traditional and locally embedded institutions with new governance mechanisms coming from the outside, thereby crafting new and hybrid institutions adapted to the socio-ecological characteristics of landscapes [2,24–26].

The distinction between these modes of governance emphasises the distinction between governance as based on clearly institutionalised central locus of authority, established rules and regulations, and a professional interpretation of the nature of the restoration process on the one hand, and governance as a process based on dispersed authority, following a flexible approach to implementation based on procedural rather than predefined ecological standards, on the other. Whereas the mode of landscape governance as a management tool is based on a refinement in the political dimension of governance, the polity and policy dimensions are not subject to major change. In contrast, the mode of landscape governance as the creation of new institutional space involves major changes in all three dimensions, as it leads to the development of new institutional arrangements at the landscape level. Such institutional bricolage [24–26] involves not only combining traditional institutions with new governance mechanisms, but also adapting nationally and internationally designed measures and plans to local circumstances. This latter form of bricolage happens when local inhabitants reject, alter or accept centrally designed rules in an attempt to maximise their positive impact, or minimise their negative impact. It also happens when policy makers decide to soften, alter or adapt the centrally designed rules in an attempt to reduce conflict with local inhabitants, or because they are familiar with local realities and realise that adaptation to local circumstances is necessary to make them fit. In both cases, such institutionally "bricoled" space is intentionally crafted to suit a landscape's socio-ecological realities better.

However, the risk is that they may lack sufficient embedding in the formal spatial decision-making structures of states, which hampers their application at larger scales [2,5].

2.3. Framework for Comparative Analysis of Cases

The various interpretations of the nature of forested landscapes and their restoration, as well as the different modes of landscape governance, have been combined into one analytical framework to allow comparative analysis of different cases of forest landscape restoration (Table 1). The table also indicates how both are related to the design principles of the landscape approach as identified by Sayer et al. [1].

Table 1. Analytical framework for assessing different interpretations of forest landscape restoration and landscape governance.

Nature of a forested landscape and its restoration	Relevant modes of landscape governance	Relationship to the main design principles formulated by Sayer et al. [1]
Ecological complex of different forest ecosystems needing restoration of ecological services	Landscape governance as a management tool	• Importance of common concern entry points as formulated in sectorial regulations and guidelines; • Strengthened stakeholder capacity for implementing professional norms
Socio-geographical space of complex mosaic land use requiring restoration of both conservation and productive functions	Landscape governance as a multi-stakeholder decision-making process	• Importance of common concern entry points deriving from multi-stakeholder negotiation process • Multi-stakeholder involvement for better coordination and planning • Negotiated and transparent change logic • Clarification of rights and responsibilities
Socio-geographical space, stretching over administrative boundaries and jurisdictions requiring restoration of both conservation and productive functions	Landscape governance as the creation of new institutional space for spatial decision making.	• multi-stakeholder involvement for joint decision making • multi-scale linkages for effective institutional embeddedness at scale • "Navigating complexity" through adaptation and continual learning

3. Research Background and Methodology

The analytical framework described in Section 2.3 served as a basis for assessing three case studies on landscape governance in Indonesia that were prepared by three MSc students from Wageningen University (Wageningen, The Netherlands) in 2012 and 2013. Each of these three students assessed the governance process behind forest landscape restoration from different angles. This section presents a systematic comparative analysis of these three studies. The analysis focuses on two main questions: (1) What form of forest landscape restoration has been at stake? (2) How was the governance process initially designed, and how did it change over time?

3.1. Research Methodology

Three of Indonesia's diverse forest restoration programmes were selected to be subjected to in-depth study (Figure 1). All three cases are part of the Masyarakat Bentang Alam Indonesia (MASBENI)—which means Landscape Community of Indonesia—a network of restoration advocates in Indonesia. All three cases have a working relationship with Tropenbos Indonesia, which is part of the Netherlands-based NGO Tropenbos International (a Netherlands-based NGO active in forest-related knowledge brokering and research [27]). The three cases were purposively selected as representing different interpretations on the nature of forested landscapes and their restoration; and representing different governance mechanisms, marked by differences in stakeholder involvement, institutional embeddedness and scale of operation. In all three cases, landscape governance has been used as a management tool, *i.e.*, as a tool to steer informal negotiations regarding managerial decisions. In only two of the cases, landscape governance has been used as a multi-stakeholder decision-making process; while in only one case landscape governance has been used to create new institutional space for spatial decision making. In view of their different geographical contexts, each of the original studies focused on location-specific issues and used specific conceptual approaches. All cases were studied through mixed methods. In each of the cases, a stakeholder analysis was carried out, based on which an average of 32 interviews were conducted among the most relevant stakeholders. This data was complemented with participatory mapping, ranking and scoring; focused discussions with mixed stakeholder groups, in-depth interviews with experts, analysis of satellite images and maps, and literature review. Further details of the precise research designs and methodologies are reported in the original studies by Hennemann [28], Brascamp [29] and Van den Dries [30], all available online. The comparative analysis of the cases presented in this article is based both on the original case study results as well as on the authors' own observations at the case study sites.

3.2. Historical Background

Indonesia is one of the countries where forest landscape restoration is high on the agenda [31,32]. The country is known for its high net loss in forest area, estimated at 8.3 million hectares from 2000–2010, representing a net decrease of about 1% per year [33]. Forest degradation, land-use conversion and fragmentation have led to a sharp reduction of ecosystem services and their benefits, which is not favourable for Indonesia's rural and urban population, nor for its economy, which is based on natural resources. Consequently, the importance of maintaining forest cover and restoring the lost forest is increasingly being acknowledged. This is reflected in the government's Green Growth Agenda, which aims to integrate ecology, economy and human welfare [34–36].

Figure 1. Location of the three case study areas in Indonesia.

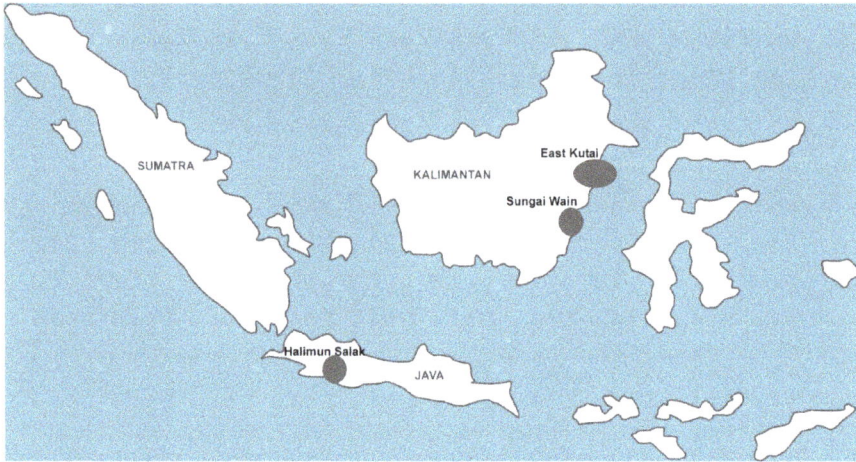

Since the second half of the 20th century, Indonesia has been a pioneer of forest landscape restoration. Initially, restoration focused on internationally sponsored watershed rehabilitation programmes. Currently, however, the scope has broadened to (urban) re-greening, restoration of waste land such as formal industrial sites, and post-mining restoration. The organisation of the restoration programmes has also gradually changed. The first watershed management programmes were managed by the Directorate of Reforestation and Land Rehabilitation, in collaboration with local communities. Currently, restoration of forested landscapes is increasingly done by governmental forestry departments in close collaboration with international conservation organisations and local NGOs, often within the framework of Reducing Emissions from Deforestation and Forest Degradation (REDD). Additionally, an increasing number of forest landscape programmes are carried out in collaboration with commercial forestry enterprises through the newly introduced ecosystem restoration concessions [36]. This latter collaboration has not always been successful. Especially during the 1990s, inappropriate incentives for encouraging timber companies to restore the timber production potential of "degraded" secondary forest resulted in the clearing of approximately 1.3 million hectares of forest land. The "degraded" sites from which previously valuable timber trees had been extracted were cleared and replanted as part of the Ministry's restoration programme [37,38]. Nonetheless, these negative experiences provided important lessons for involving commercial enterprises in forest restoration programmes in the form of industrial forest plantations. The recent shift from the restoration of forests to the restoration of landscapes, recognising the multi-functionality of forested landscapes and the variety of restoration practice, has led to new dynamism in Indonesia's forest community. A new voluntary association of landscape restoration advocates (MASBENI) has recently been formed, with the aim of actively promoting landscape restoration, in line with the international debate on integrated landscape approaches [36].

Simultaneously with the changed interpretation of forests, landscapes and their restoration, the Indonesian legal and institutional frameworks have also evolved. Whereas administrative decentralisation led to enhanced regional authority regarding the control over natural resources, including financial forest-related benefits, governmental regulation of private investments remained to be poorly monitored [39]. To allow for more transparent stakeholder involvement in forest management and restoration, new guidelines for companies investing in forest landscape restoration are currently in the making. Examples are the strict regulations for the restoration of former mining sites. Another novelty is the recognition of mosaic landscapes consisting of multiple types of land-use, in which forests provide multiple services to their inhabitants. Acknowledgement of this multi-faceted aspect of forested landscapes has led to increased inter-institutional coordination and more freedom for provincial authorities in determining the allocation of land to forestry *versus* non-forestry purposes within provincial spatial plans. There is also increased recognition of communities' multiple forest use and land rights, in an attempt to reconcile formal and informal land-use regimes. All these shifts seem to be leading to more creative restoration initiatives through multi-stakeholder arrangements at the landscape level [40].

3.3. Description of the Case Studies

The first case study was carried out in the Halimun-Salak National Park in West Java covering around 113,000 ha. This park covers the original area of Salak National Park (created in 1992), its extension towards the adjacent Halimun forest (2003), and the heavily degraded area in between. In 2003, it was proposed to restore this degraded area in between, and label it as an ecological corridor. The aim was to restore the ecological connectivity between Halimun and Salak, thus creating a much larger conservation area. Its principle focus is on restoring the landscape's original ecological structure, internal connectivity, and species mobility. An additional aim is to restore the area's function as water provider to West Java's major cities of Bogor and Jakarta. An important fact however, is that the degraded area to be restored is populated by approximately 100,000 people, who suddenly found themselves incorporated into the park, facing sharp restrictions regarding their land use and livelihood practices, which depend heavily on the natural resources (farm land, construction materials, firewood, and collection of non-timber forest products). The restoration plans therefore resulted in fierce conflicts between the inhabitants and the park's authorities [28]. To avoid further escalation, a multi-stakeholder dialogue was started, which led to the agreement that farmers can continue to farm in the area, under a number of conditions, one of which is the planting of trees. Seedlings are provided by an energy firm, operating a geothermal plant in the area.

The second case study was carried out in East Kutai District in East Kalimantan, where the private company Kaltim Prima Coal (KPC, which has a mining permit valid from 1991 until 2021) has taken the initiative to restore its former coal mining site of 90,000 ha, in line with formal government regulations. The main focus of the programme is to restore the productive function of the area, not only for commercial production, but also in the interests of the communities in and around the former mining site. These activities are based on KPC corporate social responsibility policy, which includes good post-mining management, meeting environmental standards, and involving stakeholders in the planning of social, environmental and economic development projects. Before the mining starts, the topsoil is removed and stored elsewhere. It is moved back after mining and the area is returned to its original state. This procedure is entirely in line with government regulations. KPC however has gone far beyond government regulations by initiating an intensive dialogue with local stakeholders, which has made KPC realise that just restoring the ecological structure of the forest is not enough: restoring the productive function of the landscape is more interesting to the landscape's inhabitants. KPC is therefore actively promoting a multi-functional approach to restoration, aligned with the needs and desires of the inhabitants. The costs of restoration are not covered by the company's social responsibility budget, but from the company's restoration fund, thus calculated as part of the real production costs, fully integrated in its business model [29,30].

The third case focused on the peri-urban forest of Sungai Wain, just outside Balikpapan City, East Kalimantan. Due to its proximity to the city, this 10,000 hectare forest has an important function as a provider of clean air and recreational and leisure activities for the urban people. It is also important as the major provider of clean water for the urban population and the major industries located in the area. The state-owned oil company Pertamina in particular needs large amounts of water for pumping, cooling, electricity supply and water consumption for its many employees. The area used to be heavily degraded due to fierce forest fires in the 1990s. Fire-fighting campaigns initiated from civil society resulted in massive collective action and restoration, providing Balikpapan with its current identity of a "Green, Clean and Healthy City", expressed in the Sun Bear which appears in the city's logo as well as the organisation of cultural events featuring puppet shows and songs on forest and forest restoration [30,41]. Protection of the Sungai Wain forest is still high on the local political agenda, and strict regulation mechanisms have been designed by the municipality. Forest expansion is also envisaged through the establishment of a multi-functional buffer zone, offering surrounding communities the opportunity to collect non-timber forest products and practice agro-forestry. The creation of the Botanical Garden as a tourist attraction also highlights this multi-functional approach, as it contributes to the bio-cultural identity of the area [30]. Funding for these activities is provided by the government, and the industries operating within the landscape.

4. The Results: Governing Forest Landscape Restoration in Indonesia

The three cases differ both in terms of the interpretation of forested landscapes and their form of restoration, and with respect to the mode of governance for steering decision making. However, these interpretations were gradually adjusted in all cases during the implementation of the restoration programme.

4.1. What Form of Forest Landscape Restoration Has Been at Stake?

Although all three programmes were considered as forest landscape restoration programmes, they differ significantly in their original interpretation of the nature of the forest landscape and the restoration process. Whereas two projects initially focused on restoring specific forest ecological conditions in forest reserves, the third project focused primarily on restoring the ecological services for urban residents in an urban landscape.

In Halimun-Salak, the restoration plans were initially identified by the Park Authorities in the form of an ecological corridor, devoid of agricultural activities. This plan was developed without consulting the large population (approximately 100,000) living in the area. This non-participatory approach led to serious conflict, and required adaptation of the rules: local inhabitants were allowed to farm in the newly created corridor, on the strict condition that they should actively plant trees. Notwithstanding the status as a formal conservation area, agricultural land use became tolerated as a way to mitigate conflicts and to help improve relations between governmental conservation services and local people. Consequently, local people became co-managers in the collaborative management of the forest and an energy company with local geothermal operations assisted in providing seedlings. So, while the government remained responsible for design, farmers became co-managers, and a commercial company contributed to the investments in restoration.

In East Kutai, the Kaltim Prima Coal company initially aimed to comply with the regulations of the Ministry of Mining, Energy and Mineral Resources (ESDM) regarding restoration of former mining areas; the regulations of the Ministry of Forestry regarding the structure and function of the new forest; the requirements of the Ministry of Environment for National Corporate Performance Rating Programme (PROPER); and various related regulations of the provincial and district government. However, during implementation, it was realised that establishing new forests on the denuded lands was not the primary interest of local inhabitants; hence, it was decided to broaden the scope of the restoration programme, by including community development activities (livestock rearing, agro-business and eco-tourism development, health, education and infrastructural development). In order to stimulate a process of joint planning, the original management approach was broadened to a more holistic and integrated landscape approach, with ample attention for the multi-functionality of the landscape, and the needs of local stakeholders.

In Sungai Wain, restoration activities were a direct response to the forest fires during the 1990s, and the result of collective action (NGOs, international donors and the general public). The activities did not just focus on restoring the forest cover, but rather on restoring its significance for people. The collective action provided the entire landscape with a new identity as a provider of green space and clean air for the inhabitants of Balikpapan City and clean water to Balikpapan's residents and industry. These activities contributed greatly to providing the city with a clean, green and healthy image. Within this context, the municipality has developed an active approach of involving stakeholders in formal planning procedures and implementation of management plans, while the private sector has taken care of the bulk of the investments required.

Hence, although the three projects initially differed in their interpretation of the nature of the forest landscape and the process of restoration, the interpretation of the forest landscape focused increasingly in all cases on forested mosaic landscapes.

4.2. How Was the Governance Process Initially Designed, and How Did it Change over Time?

In the cases of Halimun-Salak and East Kutai in particular, the restoration programmes were initially characterised by a professional management approach. However, during implementation there was a shift in all cases from a strict management approach to a more inclusive governance approach of stakeholder involvement. In the case of Halimun-Salak, stakeholder involvement was forced by local inhabitants supported by NGOs. Together they formed an advocacy network, and claimed institutional space to negotiate better land-use options with the Park Authorities. Thus, an informal platform was created, offering space for negotiations. An agreement was reached through this platform, allowing local people to farm within the boundaries of the extended park, but only under strict conditions. The park management realised that this would be the only way to manage the land-use conflict and create an acceptable level of co-existence [40]. In the case of East Kutai, it was KPC's initiative to involve local stakeholders, which led to a multi-functional approach to restoration. KPC recognised that involvement of local stakeholders is essential for the realisation of such a multi-functional approach; hence, KPC facilitated a platform for stakeholder participation and dialogue. Most stakeholders accepted the invitation, although some NGOs refused, as they did not agree with KPC's dominant position in the platform, and its full financial responsibility over the joint landscape design [28]. In Sungai Wain, stakeholder involvement has been strong from the onset. Born out of collective action, restoration has become high on the municipal agenda. The municipal policy is based on participatory consultation and decision making through a specially created multi-stakeholder platform, which is fully formalised [30,41]. Horizontal coordination is very strong, as governmental agencies, NGOs, industries and local communities are all represented in the Sungai Wain Protection Forest Management Body. This multi-sector management body has formal authority over the design, planning, implementation and monitoring of spatial projects.

In all cases, the process of creating institutional space has been the outcome of institutional bricolage. Not as a deliberate strategy, but as a "way in which things happen". In Halimun Salak, the bricolage was triggered by the clash between the Park Authorities and the local inhabitants, after the latter realised that the changed legal status of their land had substantial implications for their livelihoods. Through mediation of NGOs and a high level of willingness of the Park Authorities, various agreements were reached which were acceptable to both parties, yet remained informal and *ad hoc*, and recognised only for a limited period of time. In other words, the rules were bent, not changed. In East Kutai, institutional space was created by KPC, and the arrangements made were in the interests of both the company and local stakeholders. Initially, the restoration plans followed the formal government regulations, but during the process they were further adjusted and tailored to the needs of local stakeholders. During this bricolage process, local stakeholders managed to stretch the formal rules, and extended them to an outcome acceptable to all, in this case a jointly designed spatial plan. It is however not clear what the legal status of this plan is, or how it is aligned with the formal provincial planning mechanisms. The legal status of the restored land also remains unclear, which may be a source of conflict as it is unclear who will benefit from post-mining restoration, and what will happen when KPC withdraws from the area. The Sungai Wain restoration programme is clearly embedded in municipal structures and policies.

Stakeholder involvement has been formalised and embedded in the municipal administration. Here, the bricolage can be found in the way in which partners creatively used symbols and stories to gain not only political space, but also massive public support. This strong horizontal forest restoration alliance has become fully embedded in municipal politics and planning systems and is contributing greatly to the notion of the Sungai Wain forest as bio-cultural heritage contributing towards the identity of the municipality. The case shows that local-level institutional networking and bricolage is important for coherent forest landscape restoration. However, the case also shows that horizontal arrangements are not enough. Sungai Wain is currently under threat. The national government is planning to develop a new industrial area and construct the Trans Kalimantan Highway, connecting the new industrial area with the Kalimantan hinterland. This will affect Sungai Wain, as the new road is planned to pass along its border. This may result in new settlements, forest encroachment and fragmentation. Although there is strong local consent for protecting and restoring Sungai Wain, this seems to be not enough. Vertical relationships with the higher political levels are poorly developed, anchorage in national politics is weak, and economically driven decisions from higher levels overshadow local rehabilitation networks [30,42].

4.3. Overall Comparison

The analyses of the three cases indicate that their governance process differed in several respects (Table 2). In all of them, restoration programmes were initiated to serve ecological and biodiversity goals, although of a different nature. Initially, stakeholder involvement was predominantly adopted as a way to manage conflict, or to mobilise the public. Over time, however, managers became more sensitive to a more diverse set of provisioning, regulatory and cultural services of the landscape, and became more open to alternative restoration approaches better responding to the multifunctional nature of mosaic landscapes and to developing a more inclusive governance approach.

5. Discussion

Forest landscape restoration has gradually become part of the international policies on forests, climate change and food security. The understanding of its precise nature however is still developing. Forest landscape restoration is first and foremost shaped by the nature of the landscape, and the way in which the landscape is interpreted by those taking the initiative to restore. However, forest landscape restoration is also shaped by the process in which decisions are being taken regarding the aims of restoration, and the way in which restoration is implemented. This process can be referred to as landscape governance. Landscape governance differs from other forms of governance of natural resources in the sense that landscapes do not necessarily follow political or administrative boundaries, and therefore fall outside the scope of the formal spatial planning structures of states [2,5].

The emergent understanding of this multifaceted nature of forest landscape restoration is illustrated by the three Indonesian restoration programmes. The three programmes started off as a professional management approach, with the government setting the initial rules and regulations. However, over time, the rules were adapted in all three cases to the specific conditions of the landscape, and the needs and desires of the different stakeholders, evolving into a more inclusive approach of multi-stakeholder involvement. In all cases, the legal and institutional context was changed by stakeholders themselves, leading to a multi-functional approach, in which forests were placed within a wider landscape mosaic, the functions of forests were better aligned with the landscape inhabitants' needs and desires, and non-forest functions of landscapes were equally taken into account. The underlying modes of governance have stretched beyond the formal spatial planning structures and sectorial fragmentation of the Indonesian state. They have included multiple stakeholders, making them co-responsible for planning and design, but also for investing in landscape restoration. In all cases, the private sector has started to play an important role as initiator, supporter or investor in restoration [23].

In each of the three cases, flexible governance arrangements at the landscape level were lacking originally, and institutional space for negotiated decision-making at landscape level had to be claimed and created by the stakeholders involved through informal processes of bricolage [2,31]. In all cases, the formal rules were bent or changed, and turned into more flexible governance arrangements. Over time, several of these informal governance arrangements and related landscape configurations were formally recognised. This helped strengthen the landscape's identity and enhance stakeholder collaboration. In all three cases, the new governance arrangements managed to link the stakeholders into a horizontal process of spatial decisions regarding the landscape, in a more or less formalised way. Their embeddedness in the vertical or multi-layered structures of the state has however been less successful. Such embeddedness in "politics of scale" [5] seems to be a difficult yet crucial aspect of landscape governance, particularly in cases where international initiatives for forest landscape restoration require reconciliation of international, national and local interests, or in cases where landscapes are threatened by the pressures of economic development, and where stronger resilience of landscapes is needed in the face of externally driven resource exploitation and infrastructural development.

Table 2. Comparative overview of the governance process of three cases of forest landscape restoration in Indonesia.

Case study	Original restoration approach	Mode of governance	Evolution in governance approach
Halimun-Salak	Restoration of an area degraded due to agricultural expansion. Restoration of an ecological corridor to restore ecological integrity and species mobility	Landscape governance as a management tool: plans are designed and implemented by Park Authorities; stakeholder involvement merely seen as a conflict management tool	Initially not participatory and highly directive. However, focus changed to more stakeholder involvement to mitigate conflict. Multiple resource use negotiated and accepted, yet not legalised. Institutional space claimed by local inhabitants with NGO support, but not institutionalised. Main funder: government. Additional funding provided by private sector
East Kutai (KPC)	Restoration of former mining sites, emphasis on restoring the original forest cover	landscape governance as a multi-stakeholder decision-making process: within the formal government regulations on restoration there is room for multi-stakeholder dialogue, which has led to more creative multifunctional restoration practice (agriculture, livestock, tourism)	Initially focused on implementation of government regulation, but later on turned into an instrument for participatory spatial planning. Institutional space created for multiple land use. Institutional space created by the company, in agreement with a majority of local stakeholders, yet not formalised or institutionalised in formal planning mechanisms of the government. Main funding: private sector
Sungai Wain	Restoration of fire damage. Emphasis on ecological restoration, provision of clean water and cultural identity	Landscape governance as the creation of new institutional space for spatial decision making; collective action and strong multi-stakeholder collaboration has led to new space for decision making, institutionalised in local government authorities	Integrated and multi-stakeholder approach from the onset; stakeholder involvement as instrument for joint planning; institutional space for multi-stakeholder dialogue created, and formally embedded in local government and its planning mechanism, however poorly embedded in national politics. Main funding: initially civil society and international donors. Later on: municipal government, with substantial co-funding from industries operating in the area

6. Conclusions

Our analysis indicates that forest landscape restoration should not be based only on design criteria such as formulated by Sayer *et al.* [1], but rather on a good understanding of (a) the different interpretations of the substantive nature of forest landscapes and their restoration needs; and (b) the different modes of landscape governance including the dynamics of their institutionalisation. Our analysis underlines the opinions of various authors [2,4,5] that forest landscape restoration must be based on the notion that local realities matter. It emphasises that landscape restoration requires a flexible approach of social learning rather than a strongly institutionalised approach based on design criteria. To be successful, also landscape governance has to be based on a thorough understanding of the nature of forest landscapes and their restoration. It cannot be solely based on considerations of the political dimensions of governance (with special attention to the participation of non-state organisations and private actors), but must include considerations on how best to incorporate space for social learning and a gradual adaptation of the polity and policy dimensions of governance through a process of institutional bricolage. All landscapes are fundamentally different, as they are the product of socio-ecological processes that are unique in time and place. It is therefore not only important to assess global potentials and design globally applicable instruments and guidelines, but also to support local landscape's stakeholders in planning and designing their own restoration programmes according to their specific needs and, more importantly, to help develop multi-actor, multi-sector and multi-scaled governance mechanisms that allow locally designed plans to be linked to overall planning mechanisms of the state. Most importantly of all, it has to be accepted that forest landscape restoration cannot be based on professional design alone, but rather depends on gradual changes in both the substance and the modes of governance, which emerge out of local creativity and the gradual emergence of innovative public–private arrangements at the landscape level.

Acknowledgments

The authors would like to acknowledge the work of Lien Imbrechts [43], Ilse Hennemann [28], Fenneke Brascamp [29] and Bas van den Dries [30], the MSc students who collected the original data that were used to compile this comparative analysis. The authors also wish to express their gratitude to the management of Halimun Salak National Park, the management and staff of Kaltim Prima Coal, and the government authorities of Balikpapan City, who hosted the students while carrying out their research. Final thanks go to all the communities and individuals living in and around the Halimun Salak National Park, East Kutai and Balikpapan City, who so actively participated in the field studies and observations. The research would not have been possible without them. The authors also acknowledge three anonymous reviewers whose comments helped develop the paper.

Author Contributions

Cora van Oosten and Freerk Wiersum are responsible for the overall research design and supervision of the entire research process. They are the main authors of the analytical framework, and the framework for the comparative analysis of the cases. Petrus Gunarso and Irene Koesoetjahjo are responsible for the field work, and supervised the process of data collection in Indonesia. They are the main authors of the description of the description of the research background, the historical background, and the case studies. All the authors are co-responsible for the analysis of the results, the discussion and the final conclusion.

Conflicts of Interest

The authors declare no conflict of interest.

References

1. Sayer, J.; Sunderland, T.; Ghasoulc, J.; Pfund, J.L.; Sheilb, D.; Meijaard, E.; Ventera, M.; Boedhihartonoa, A.K.; Day, M.; Garcia, C.; *et al.* Ten principles for a landscape approach to reconciling agriculture, conservation, and other competing land-uses. *PNAS* **2013**, *110*, 8349–8356.
2. Van Oosten, C.J. Restoring landscapes—Governing place: A learning approach to forest landscape restoration. *J. Sustain. For.* **2013**, *32*, 659–676.
3. Global Partnership on Forest and Landscape Restoration, 2011. Available online: http://www.forestlandscaperestoration.org (accessed on 7 November 2013).
4. Colfer, C.J.P. *Collaborative Governance of Tropical Landscapes*; Earthscan: London, UK, 2011.
5. Görg, C. Landscape governance: The "politics of scale" and the "natural" conditions of places. *Geoforum* **2007**, *38*, 954–966.
6. Kooiman, J. *Governing as Governance*; Sage: London, UK, 2003.
7. Kooiman, J. Governability: A conceptual exploration. *J. Comp. Policy Anal.* **2008**, *10*, 171–190.
8. Hamilton, L.S. *Forest and Watershed Development and Conservation in Asia and the Pacific*; Westview Press: New York, NY, USA, 1983.
9. Easter, K.W.; Dixon, J.A.; Hufschmidt, M.M. *Watershed Resources Management. Studies from Asia and the Pacific*; Westview Press: New York, NY, USA, 1986.
10. Savenije H.; Huijsman, A. *Making Haste Slowly, Strengthening Local Environmental Management in Agricultural Development*; Development Oriented Research in Agriculture, Royal Tropical Institute: Amsterdam, The Netherlands, 1991; Volume 2.
11. Bishop, J.; Landell-Mills, M. Forest environmental services: An overview. In *Selling Forest Environmental Services: Market-Based Mechanisms for Conservation and Development*; Pagioli, S., Bishop, J., Landell-Mills, M., Eds.; Earthscan Publications: London, UK, 2002; pp. 15–36.
12. Reid, W.V.; Mooney, H.A.; Cropper, A.; Capistrano, D.; Carpenter, S.; Chopra, K.; Dasgupta, P.; Dietz, T.; Duraiappah, A.K.; Hassan, R.; *et al. Millennium Ecosystem Assessment—Ecosystems and Human Well-Being: Current State and Trends*; Island Press: Washington, DC, USA, 2005; Volume 1, pp. 25–64.

13. Wiersum, K.F. Forest dynamics in the tropics. In *Forestry in a Global Context*, 2nd ed.; Sands, R., Ed.; Commonwealth Agricultural Bureaux International Publishing (CAB): Wallingford, UK, 2013; pp. 119–132.

14. Wiersum, K.F. From natural forest to tree crops, co-domestication of forest and tree species, an overview. *Neth. J. Agric. Sci.* **1997**, *445*, 425–438.

15. Michon, G.; de Foresta, H.; Levang, P.; Verdeaux, F. Domestic forests: A new paradigm for integrating local communities into tropical forest science. *Ecol. Soc.* **2007**, *12*, 1.

16. Genin, D.; Aumeeruddy-Thomas, Y.; Balent, G.; Nasi, R. The multiple dimensions of rural forests: Lessons from a comparative analysis. *Ecol. Soc.* **2013**, *18*, 27.

17. Lemenih, M.; Wiersum, K.F.; Woldeamanuel, T.; Bongers, F. Diversity and dynamics of management of gum and resin resources in Ethiopia: A trade-off between domestication and degradation. In *Land Degradation & Development*; John Wiley & Sons, Ltd.: Hoboken, NJ, USA, 2011; doi:10.1002/ldr.1153.

18. McKey, D.; Linares, O.F.; Clement, C.R.; Hladik, C.M. Evolution and history of tropical forests in relation to food availability—Background. In *Tropical Forests, People and Food. Biocultural Interactions and Application to Development*; Hladik, C.M., Hladik, A., Linares, O.F., Pagezy, H., Semple, A., Hadley, M., Eds.; Man and Biosphere Series 13; Partenon United Nations Educational, Scientific and Cultural Organization: New York, NY, USA, 1993; pp. 17–24.

19. Hecht, S.B.; Padoch, C.; Morrison, K. *The Social Lives of Forests: Woodland Resurgence and Forest Landscapes in the Past, Present and Future*; University of Chicago Press: Chicago, IL, USA, 2010.

20. Chomitz, K.M. *At Loggerheads? Agricultural Expansion, Poverty Reduction, and Environment in Tropical Forests*; World Bank: Washington, DC, USA, 2007.

21. Arts, B.; Visseren-Hamakers, I. Forest governance: A state of the art review. In *Forest-People Interactions: Understanding Community Forestry and Biocultural Diversity*; Arts, B., van Bommel, S., Ros-Tonen, M., Verschoor, G., Eds.; Wageningen Academic Publishers: Wageningen, The Netherlands, 2012; pp. 241–257.

22. Treib, O.; Bähr, H.; Falkner, G. Modes of governance—Towards a conceptual clarification. *J. Eur. Public Policy* **2007**, *14*, 1–20.

23. Van Noordwijk, M.; Tomich, T.P.; Verbist, P. Negotiation support models for integrated natural resource management in tropical forest margins. In *Integrated Natural Resource Management. Linking Productivity, the Environment and Development*; Campbell, B.M., Sayer, J.A., Eds.; Commonwealth Agricultural Bureaux International Publishing (CABI): Wallingford, UK, 2003; pp. 87–108.

24. Cleaver, F. Reinventing institutions: Bricolage and the social embeddedness of natural resource management. *Eur. J. Dev. Res.* **2002**, *14*, 11–30.

25. Cleaver, F. *Development through Bricolage: Rethinking Institutions for Natural Resource Management*; Routledge: New York, NY, USA, 2012.

26. De Koning, J.; Cleaver, F. Institutional bricolage in community forestry: An agenda for future research. In *Forest-People Interactions: Understanding Community Forestry and Biocultural Diversity*; Arts, B., van Bommel, S., Ros-Tonen, M., Verschoor, G., Eds.; Wageningen Academic Publishers: Wageningen, the Netherlands, 2012; pp. 277–290.

27. Tropenbos International—Making Knowledge Work for Forests and People. Available online: http://www.tropenbos.org/ (accessed on 10 December 2013).

28. Hennemann, I. Looking beyond the forest: Exploring the significance of alignment between politics and practices in landscape governance. A case study of the corridor in Gunung Halimun Salak National Park in West Java, Indonesia, 2012. Wageningen University, Environmental Policy Group. Available online: http://edepot.wur.nl/244457 (accessed on 10 December 2013).

29. Brascamp, F. Landscape restoration through innovative landscape governance. A case study of coal mining in East Kalimantan, Indonesia, 2013. Wageningen University & Research, Forest and Nature Conservation Policy Group. Available online: http://www.forestlandscaperestoration/org/resource/landscape-estoration-through-innovative-landscape-governance-case-study-coal-mining-east-ka (accessed on 3 May 2014).

30. van den Dries, B. The Landscape Governance Process for Conserving the Sungai Wain Protection Forest. Master's Thesis, Wageningen University, Wageningen, The Netherlands, August 2013. Available online: http://www.forestlandscaperestoration.org/resource/landscape-governance-process-conserving-sungai-wain-protection-forest-indonesia (accessed on 3 May 2014).

31. Nawir, A.A.; Murniati; Rumboko, L.; Gumartini, T. The historical national overview and characteristics of rehabilitation initiatives. In *Forest Rehabilitation in Indonesia: Where to After More Than Three Decades?* Nawir, A.A., Murniati, Rumboko, L., Eds.; Center for International Forestry Research (CIFOR): Bogor, Indonesia, 2007; Chapter 4.

32. Nawir, A.A.; Murniati; Rumboko, L.; Hiyama, C.; Gumartini, T. Past and present policies and program memes affecting forest and land rehabilitation initiatives. In *Forest Rehabilitation in Indonesia: Where to After More Than Three Decades?* Nawir, A.A., Murniati, Rumboko, L., Eds.; Center for International Forestry Research (CIFOR): Bogor, Indonesia, 2007; Chapter 3.

33. Miettinen, J.; Shi, C.H.; Liew, S.C. Deforestation rates in insular Southeast Asia between 2000 and 2010. In *Global Change Biology*; Blackwell Publishing: Hoboken, NJ, USA, 2011; Volume 17, pp. 2261–2270.

34. Badan Kebijakan Fiskal. *Indonesia's Green Growth Strategy for Global Initiatives: Developing A Simple Model and Indicators of Green Fiscal Policy in Indonesia*; Badan Kebijakan Fiskal: Jakarta, Indonesia, 2011.

35. Ministry of Forestry. Rehabilitasi lahan dan perhutanan sosial (Land rehabilitation and social forestry), Statistik Kehutanan Indonesia 2001 (Forestry statistics Indonesia 2001) Ministry of Forestry (MoF), The Government of Indonesia, Jakarta. Available online: http://www.dephut.go.id/INFORMASI/STATISTIK/2001/RLP_01_N.htm (accessed on 23 February 2008).

36. Gunarso, P. National Workshop on Systematisation of Forest Productivity Improvement through Landscape Restoration Programme in Indonesia, Mataram, Lombok, Indonesia, 5–6 December 2013.

37. Barr, C. HPH timber concession reform: Questioning the "sustainable logging" paradigm. In *Which Way Forward? People, Forests, and Policymaking in Indonesia*; Colfer, C.J.P., Resosudarmo, I.A.P., Eds.; Resources for the Future: Washington, DC, USA, 2002.

38. Barr, C.M.; Sayer, J.A. The political economy of reforestation and forest restoration in Asia-Pacific: Critical issues for REDD+. Special Issue Article: REDD+ and conservation. *Biol. Conserv.* **2012**, *154*, 9–19.

39. Obidzinski, K.; Barr, C. *The Effects of Decentralisation on Forests and Forest Industries in Berau District, East Kalimantan*; Center for International Forestry Research (CIFOR): Bogor, Indonesia, 2003.

40. Royo, N.; Wells, A. *Community Based Forest Management in Indonesia: A Review of Current Practice and Regulatory Frameworks*; Investing in Locally Controlled Forestry (ILCF): Yogyakarta, Indonesia, 6–9 February 2012.

41. Kaltim Prima Coal. Sustainability report 2010 expansion for sustainability, 2010. Kaltim Prima Coal. Available online: http://www.kpc.co.id/ (accessed on 12 February 2012).

42. Fredriksson, G.M.; de Kam, M. Strategic plan for the conservation of the Sungai Wain Protection Forest, East Kalimantan, Indonesia. In *The International Ministry of Forestry and Estate Crops—Tropenbos Kalimantan Project*; Tropenbos Kalimantan Project: Balikpapan, Indonesia, 1999; pp. 1–38.

43. Imbrechts, L. Protection Forest Governance: Inside-out: Exploring the Gaps between Discourse and Reality. Master's Thesis, Katholieke Universiteit Leuven, Leuven, Belgium, November 2011. Available online: http://www.scriptiebank.be/sites/default/files/88905cfc8ad 8961e011ab5c400254450.pdf (accessed on 3 September 2012).

Challenges of Governing Second-Growth Forests: A Case Study from the Brazilian Amazonian State of Pará

Ima Célia Guimarães Vieira, Toby Gardner, Joice Ferreira, Alexander C. Lees and Jos Barlow

Abstract: Despite the growing ecological and social importance of second-growth and regenerating forests across much of the world, significant inconsistencies remain in the legal framework governing these forests in many tropical countries and elsewhere. Such inconsistencies and uncertainties undermine attempts to improve both the transparency and sustainability of management regimes. Here, we present a case-study overview of some of the main challenges facing the governance of second-growth forests and the forest restoration process in the Brazilian Amazon, with a focus on the state of Pará, which is both the most populous state in the Amazon and the state with the highest rates of deforestation in recent years. First, we briefly review the history of environmental governance in Brazil that has led to the current system of legislation governing second-growth forests and the forest restoration process in Pará. Next, we draw on this review to examine the kinds of legislative and operational impediments that stand in the way of the development and implementation of a more effective governance system. In particular, we highlight problems created by significant ambiguities in legal terminology and inconsistencies in guidance given across different levels of government. We also outline some persistent problems with the implementation of legal guidance, including the need to understand local biophysical factors in order to guide an effective restoration program, as well as difficulties presented by access to technical assistance, institutional support and financial resources for the establishment and monitoring of both existing secondary forests and newly regenerating areas of forest. Whilst we focus here on a Brazilian case study, we suggest that these kinds of impediments to the good governance of second-growth forests are commonplace and require more concerted attention from researchers, managers and policy makers.

Reprinted from *Forests*. Cite as: Vieira, I.C.G.; Gardner, T.; Ferreira, J.; Lees, A.C.; Barlow, J. Challenges of Governing Second-Growth Forests: A Case Study from the Brazilian Amazonian State of Pará. *Forests* **2014**, *5*, 1737-1752.

1. Introduction

Second-growth forests (*i.e.*, forests regenerating on areas that have previously been clear-cut) are an increasingly ubiquitous element of human-modified landscapes and currently account for more than half of the world's remaining tropical moist forests [1,2]. These forests can provide critically important habitat for safeguarding biodiversity, especially in parts of the world where native vegetation is highly fragmented or where there is little old-growth forest remaining [3–5]. Second-growth forests can also provide significant ecosystem services, including the recovery of soil fertility in fallow farming systems [6,7], the provision of natural resources to support local livelihoods [8] and carbon sequestration and conservation [9–12].

Yet, second-growth forests can often be highly ephemeral components of a landscape [13]. Many human-modified landscapes in agricultural frontier regions are comprised of complex and dynamic patchworks of agricultural areas and fragments of regenerating forest [14]. The fate of a given patch of second-growth forest is determined by the interplay between economic incentives for returning the land to production, the value of the forest to local people (including as fallow land for farmers who lack access to external sources of nutrients) and the legal framework governing the management and clearance of such forests [15]. Yet, the legal framework governing second-growth forests in many countries and especially in those with active deforestation frontiers is frequently marked by high levels of uncertainty and controversy. This is for at least two main reasons. First, the fact that such landscapes are highly dynamic, with shifting patterns of active agricultural production, fallow and land abandonment, makes it hard to design, implement and monitor any regulations on second-growth management and clearance practices. Second, the value of second-growth forests to society is often poorly appreciated, and they are commonly viewed as areas of degraded land with little or no economic value. This perception is exacerbated by considerable uncertainty and disagreement regarding the point at which a forest regenerating on once-cleared land can legally be classified as a "forest".

Here, we present an overview of some of the particular challenges facing the governance of second-growth forests. We use the state of Pará, in the eastern Brazilian Amazon, as a case study, as it is typical of many agricultural frontier regions across the tropics. Significant agricultural land abandonment has been observed in this region since 1940 [16], and about 25% of the deforested area was under some form of second-growth forest in 2010 [17]. Forty percent of this was in the state of Pará, where the Brazilian government's TerraClass mapping program identified over 165,000 km^2 of second-growth vegetation in 2010 (Figure 1). Although some regenerating stands in older landscapes, such as in the northeast of Pará, can be over 50 years old, most second-growth forests in the Brazilian Amazon, as elsewhere, are relatively short-lived components of a landscape, with an average age of only five years in 2002 [18].

To provide context, we first briefly examine the legal framework governing environmental resources and second-growth forests across Brazil, before focusing on evolving governance structures of the state of Pará. We then analyse some of the key impediments, including both legal and implementation aspects, facing the development of a system of good governance for second-growth forests. In discussing second-growth forests, we are concerned with both the governance of established areas of second-growth forest, as well as the governance, whether through passive or active approaches to restoration, of forests regenerating on cleared land. Forest restoration is of particular prominence in Brazil following the recent revision and renewed enforcement of the Forest Code (Código Florestal)—the central piece of legislation regulating land use and management on private properties. We end the paper with a brief discussion on how many of these issues are generic to agricultural regions across the world.

Figure 1. Distribution of secondary forests in Pará state, Brazil. Source: Instituto Nacional de Pesquisas Espaciais—Centro Regional da Amazônia (CRA/INPE).

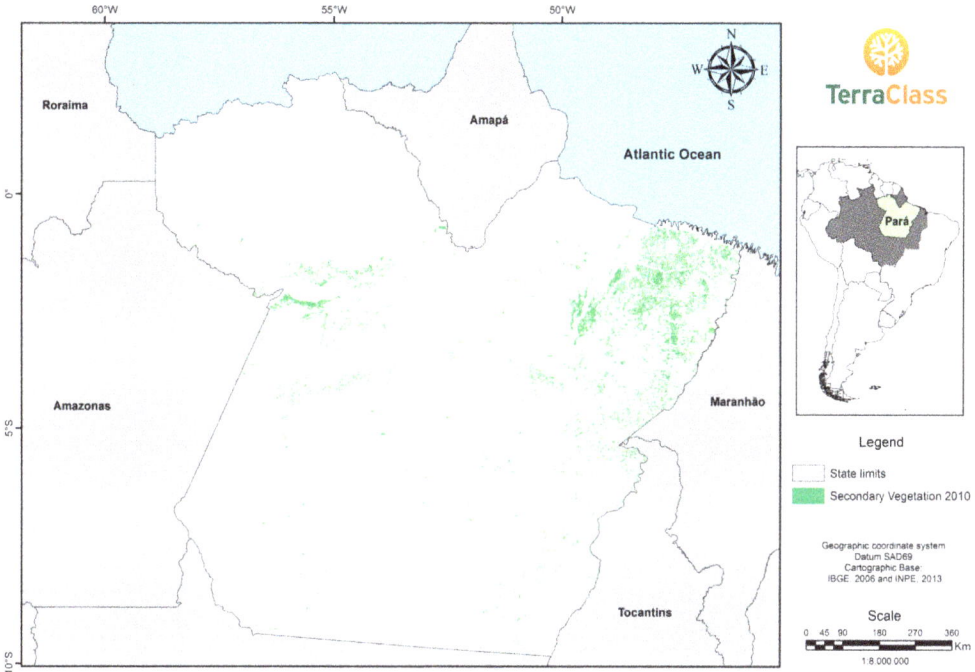

2. A Historical Perspective of the Governance of Second-Growth Forests and Forest Restoration in Brazil and the State of Pará

Across Brazil, many aspects of the legal framework governing second-growth forests and the regeneration of degraded land remain both poorly clarified and understood [19,20]. Persistent ambiguities and uncertainties in the legal framework governing second-growth forests in Brazil has led to a lack of consistency and coherence between different levels of environmental governance, as well as negatively affecting the perception of their importance to society by both the agricultural sector and legislators alike. Part of the confusion relates to the distinctly hierarchical nature of Brazils' federal governance system, where regulations imposed by states and municipalities cannot be seen to undermine federal directives (*i.e.*, they can only be more, not less, environmentally conservative). However, in practice, the state level, through the actions of state environmental secretaries, often emerges as the dominant player in legislating, regulating and controlling environmental impacts.

To better understand these complexities, it is necessary to evaluate the history of second-growth forests (and regenerating areas of other, non-forest ecosystems) that have been recognized by Brazilian law. In 1981, the National Environment Policy of Brazil (Law No. 6938) highlighted the "restoration of degraded areas" as a national priority, including for mitigation and compensation activities related to development impacts (e.g., infrastructure, mining and oil and gas). Subsequent to this, the Environmental Criminal Law (1998, Law No. 9605) gave further legal weight to the importance of restoration and the value of second-growth forests, by stipulating that the restoration of degraded areas should form an obligatory part of environmental mitigation strategies.

However, despite the recognition given to second-growth forests in these early pieces of environmental legislation, their significance in the eyes of many legislators and environmental implementing agencies continued to remain limited, especially in the agricultural sector. This situation changed abruptly following the revision of the Forest Code in 2012 (Law No. 12651/2012), which attracted enormous national and international attention and heralded the start of a new phase in the development and implementation of forest policy in Brazil [20]. After multiple vetoes and revisions, the revised law finally came into force in October of the same year (Presidential Decree 12727/12).

The revised Forest Code introduced new mechanisms and criteria to determine the areas that need to be conserved and/or regenerated, depending on property size and the length of time since the original forest was cleared. A central pillar to its implementation is the Cadastro Ambiental Rural (CAR) (Table 1), an electronic land registration system, where landowners declare the current legal status of each rural property in terms of legal reserves (LR), areas of permanent protection (APP) and production areas.

Considering the entire country, the new law identified some 21 million hectares of illegally cleared land that must be restored, of which 78% is in private legal reserves and the remainder in areas of permanent protection, including areas of riparian vegetation and other environmentally sensitive areas, such as steep slopes and hilltops [21]. With respect to LRs in properties, where the CAR system shows a reserve deficit (which is less than 80% in forested areas of the legal Amazon) prior to July 22, 2008, the deficit area must be restored within 20 years. By contrast, properties with a deficit created after July 22, 2008, are obliged to immediately suspend all production activities, and regeneration strategies must be put in place by 2014. In the case of APPs, areas that were already consolidated for agriculture prior to July 22, 2008, are "allowed to continue in production, provided that, and depending on property size and other factors, minimum-sized areas are regenerated". Under the revised Forest Code, the total area of environmental deficit, *i.e.*, areas deforested prior to 2008 that must now be restored, has been reduced by 58% compared to the original law, mostly due to an amnesty given to smaller properties, permitting the inclusion of APPs in the calculation of the total LR area, a reduction in the LR restoration requirement to 50% in municipalities dominated by protected areas and relaxing of the restoration requirements in APPs on smaller properties and those designated for agricultural production [20].

The changes and adaptations to the federal Forest Code generated an enormous legislative and implementation burden for individual states, which were passed the responsibility of prescribing specific laws and regulations for enforcing forest governance at the regional level. For example, properties with a deficit of legal reserve or APP must adopt regeneration strategies in agreement with the relevant state's Program for Environmental Regularization (Programa de Regularização Ambiental, PRA) (Table 1). For the majority of states, this is underpinned by the Rural Activities License (LAR), which acts as the main regulatory mechanism for the overall organization of land use, including any areas set aside for restoration. The LAR sets a legal obligation to bring the property into compliance with environmental regulation, including the restoration of or compensation of LRs and APPs (Table 1). Beyond the establishment of state-level PRAs to guide the restoration process for illegally deforested areas, state-level regulations are also needed to prescribe the governance of existing second-growth forests and whether they should be classified as either forest or fallow land. For example, in many states, there is no legal definition on when a regenerating area becomes classified as "forest" compared to "fallow" and qualifies for legal protection. There is also a lack of guidance on the ways in which fallow areas can be cleared, as well as the restoration techniques that should be used, whether active or passive, to restore land that was illegally deforested in the past, despite an emerging body of literature on the subject [22]. In combination, such legislative shortcomings often open the door for *ad hoc* and inconsistent decision-making that, in the long-run, is likely to severely undermine this undervalued environmental resource.

Currently, Pará is the only state of the Brazilian Amazon that has adopted an explicit definition of second-growth forests, *i.e.*, forests that have regenerated from previously cleared land and that can no longer be considered as fallow (and, hence, cannot be cleared). For now, the state has defined a lag time of three years before any active restoration activity is initiated in order to evaluate the potential for passive restoration (thereby reducing costs). This delay period was established in recognition of the natural propensity for the regeneration in the Amazon region if the previous land use had not been excessively intense [22]. The regulations proposed in the PRA further stipulate that any necessary active restoration activities must be completed within nine years for APPs and 20 years for LRs, with frequent periods of monitoring for each.

In February, 2014, Pará established the first legislation defining successional stages of second-growth forests for any Amazonian state (Instrução Normativa, IN 02 February 26 2014). Prior to establishing this law, the management of different types of second-growth forests and their clearance for agriculture represented something of a legal vacuum, preventing the implementation of state zoning legislation (State Decree 7398, 2010), which stipulates that intermediate and advanced-stage second-growth forests should be conserved. Following the passing of IN 02 February 26 2014, licenses for the clearance of second-growth forest in a given private property must be based on a combination of age, the basal area of large trees and the percentage of primary forest in the municipality where it is located. Stands of second-growth forest shown through inspection of satellite images to be older than 20 years are recommended for protection without requiring any field assessment, while stands between five and 20 years old are recommended for protection, depending on the total stand basal area of native trees and palms equal to or larger than 10 cm in diameter. The threshold to authorise the clearance for agriculture is less than 10 m^2 ha^{-1} for municipalities with higher than 50% primary forest cover and less than 5 m^2 ha^{-1} for those municipalities with less than 50% primary forests.

Table 1. Legal instruments associated with the Brazilian Forest Code (Law 12651/2012) and its relation to the governance of second-growth forests and the forest restoration process in Brazil.

Legal Instruments	Definition	Role in the Restoration Process
Permanent Protection Areas (APPs)	Environmentally sensitive areas that must be legally protected to conserve water resources, geological stability, biodiversity, facilitate gene flow of fauna and flora, afford soil protection and support the well-being of local human populations. APPs include both riparian areas that protect riverside forest, hilltops and steep slopes.	The amount of APP that must be restored is proportional to the size of the property, as well as the water body in question in the case of riparian areas. Property size (fiscal module) — Width of APP to be restored (m) Up to 1 — 5, for all water courses 1 to 2 — 8, for all water courses 2 to 4 — 15, for all water courses 4 to 10 — 20 to 100, depending on water course width Above 10 — 30 to 100, depending on water course width All — 15, for all perennial springs.
Legal Reserves (LRs)	Legal reserves are fixed minimum percentages of properties that must be left as native vegetation, the size of which varies by biome. Some sustainable use is allowed (such as selective logging), but the areas must assist in the conservation and rehabilitation of ecological processes and promote biodiversity conservation.	Following the revised Forest Code, restoration of LRs must take place in 20 years or less. This can be achieved passively through natural regeneration or by planting different native species. The use of fruit, ornamental or exotic cultivated species is also allowed in small (family) farms, when intercropped with native tree species.
SICAR (Sistema de Cadastro Ambiental Rural)	The SICAR is a nationwide electronic system for the management of environmental information of rural properties	SICAR permits the federal government to more effectively oversee land use planning and environmental conservation initiatives nationwide, as well as allowing individual property owners to assess their own compliance with the revised Forest Code. From 2017, financial institutions will only grant credit to CAR-registered rural landowners.

Table 1. *Cont.*

Legal Instruments	Definition	Role in the Restoration Process
Rural Activities License: LAR	The main regulatory mechanism for the overall organization of land use, including the restoration process established by Pará state decree (857, January 30, 2004, and subsequently edited (Decree 216, September 22, 2011).	Determines what agricultural activities can be conducted in areas where native vegetation has already been modified/cleared and represents the main instrument for controlling, monitoring and demonstrating the environmental compliance of private properties in Pará, especially regarding the management of legal reserves and permanent protection areas.
Program for Environmental Regularization (PRA)	The Programa de Regularização Ambiental (PRA) is the state-level instrument used to implement the Presidential Decree 8235 with a set of actions or initiatives to be undertaken by rural landowners and leaseholders in order to achieve environmental compliance.	Deals with the regularization of APPs and LRs and restricted use (RU) upon recovery, restoration, regeneration or compensation. The owners or occupiers of rural properties should hold the PRA after completing the Rural Environmental Registry (CAR). The decree complements the rules necessary for the implementation of the CAR, which will start the process of rural environmental restoration planned in the current Forest Code.

The above criteria, developed through a scientific advisory working group (including the authors of this article) under the auspices of the state environmental department, SEMA and the Programa Municípios Verdes (a state-wide program to reduce deforestation and promote the adoption of more sustainable land-use systems), took into consideration the recovery of biodiversity and ecosystem service provision using field information from 140 forest plots across the state.

The definition of second-growth successional stages also informs how these areas can be used for achieving environmental compliance. For example, according to the current proposal for the state PRA, areas can only be traded if they are in an intermediate or advanced stage of regeneration (although areas in an initial stage of regeneration can be rented to a third party through a strictly bilateral arrangement). This focus on the older second-growth forests helps avoid inundating the market for the trading of legal reserve credits to achieve compliance with the Forest Code with vast areas of very young second-growth forest.

3. Impediments and Challenges to the Good Governance of Second-Growth Forests and the Restoration Process in the State of Pará

Despite the advances made by the Brazilian federal- and state-level environmental legislation outlined above, there are many challenges involved in ensuring that these changes result in the good governance of second-growth forests, *i.e.*, safeguarding the long-term protection of the environmental services provided by these forests, whilst ensuring the fair and sustainable development of the agricultural sector, including the need for adequate resources and technical assistance to support the regeneration of degraded and illegally deforested areas. We outline some of these challenges using the state of Pará as our case study, first assessing some of the key legal impediments to good governance and then examining the operational challenges, highlighting four key issues that complicate translating legal prescriptions into practice. Our assessment is far from exhaustive and reflects our own experiences of some of the key barriers to the effective conservation of existing second-growth forests and the large-scale restoration of degraded areas in Pará. However, we believe that many of these issues are generic to the conservation and restoration of second-growth forests in other Brazilian states and other nations that host tropical forests. By highlighting these persistent problems, we hope to contribute towards efforts to develop more clear, consistent and fair regulatory frameworks for second-growth forests across the tropics.

3.1. Legal Impediments to Achieving Good Governance of Second-Growth Forests

The legislative frameworks governing second-growth forests and the process of restoring forest on illegally cleared land in the state of Pará have a wide range of short-comings that are not limited to this state and are symptomatic of widespread difficulties in developing clear, consistent and fair rules for the management of second-growth forests, and indeed, environmental resources in general, in many parts of the world. Here, we briefly discuss four types of impediment that can be observed in Pará today and exemplify some of the problems that they generate, including a lack of clarity in key definitions; inconsistencies in legal frameworks between different levels of governance and over time; and the potential for abuse or "loopholes" in how regulations are enforced.

3.1.1. Unclear or Poorly Founded Definitions and Concepts

An important conceptual challenge facing the governance of second-growth forests in Brazil in general is the diversity of scientific and technical terms used to describe key issues, such as the process of forest restoration itself, and a lack of consensus regarding the use of these terms amongst different actors. For example, the revised Forest Code refers to the obligation to restore native vegetation where private properties are not compliant with the law, but uses a variety of terms to describe this, including forest recovery, restoration and recuperation. Restoration is considered by this law to be the "recovery of a degraded ecosystem or a wild population to a state as close as possible to its original condition", while recuperation is the "recovery of a degraded ecosystem or a wild population to a non-degraded condition, which may be different from its original condition", while the term recovery refers more broadly to land, soil, vegetation and the environment generally. The use of these contrasting terms often interchangeably can generate confusion as to the overarching aims of a given piece of legislation.

In addition, regulations that prescribe how second-growth forests should be protected, cleared and managed are often made without any clear justification or supporting evidence. For example, prior to the approval of the new law (IN 02/2014), the interim legislation decreed that second-growth forests could be cleared for agriculture wherever there was a density of less than 50 trees larger than 10 cm DBH (diameter at breast height) per hectare. However, it has been impossible to find any documentation or evidence supporting why this "50 trees per hectare" rule was chosen.

3.1.2. Inconsistencies between Different Levels of the Legal Framework

The responsibility for the environmental governance of private land in Brazil is distributed across federal, state and municipal levels. However, delays in the specification of general frameworks provided by the federal government at state and municipal levels, as well as differences in priorities between different levels of government can commonly result in uncertainty and contradictions regarding the interpretation of the law at the local level. For example, there is a lack of specific guidance on the techniques that should be used to facilitate regeneration in areas that must be restored by law (e.g., deforested riparian zones), such as the type (e.g., native or non-native) and number of species that should be used, resulting in the potential for varying interpretations and possible development-conservation conflicts [23]. A wider problem in aligning the requirements imposed by federal legislation with implementation at the local level is the need for political continuity in the state government, which is primarily responsible for managing this process. This is especially problematic in the case of long-term environmental problems, such as the regeneration of second-growth forests that play out over multiple election periods. The failure of the "1 billion trees" restoration program initiated by the previous Pará state government provides a clear example of this. This program developed the first technical guidance for restoration projects in Pará (State Decree 1848, August 21, 2009), but despite its political importance, it suffered from significant strategic and operational problems and was discontinued two years after implementation.

3.1.3. Frequent Revisions to Legal Documents over Time

A key requirement for a given piece of legislation to be regulated and implemented in practice is that it remains stable for a minimum period of time. However, in practice, Brazilian environmental legislation is commonly characterized by frequent revisions and alterations. The federal Forest Code is perhaps the most famous example of this, which was subject to a very large number of provisional amendments before the current version was finally agreed upon in October, 2012. In the state of Pará, the main piece of legislation governing the conservation and restoration of forests in private properties was altered three times in four years (State Decrees, 2141 in March, 2006, 1848 in August, 2009, and 2099 in January, 2010), with each new revision revoking the authority of the previous version. These changes generated significant controversy and uncertainty and prevented the finalization of specific regulations (e.g., the size of legal reserves in areas with state zoning legislation) that are needed to implement any new law once it has been approved. The personal experience of the authors is that extremely few individuals, whether legislators, enforcers or landowners, have a strong command of the full set of legal prescriptions for the management and clearance of second-growth forests at any given point in time. This situation is further exacerbated by the frequent turn-over of key individuals in state- and municipality-level government.

3.1.4. Potential for Abuse and the Inequitable Application of Regulations

A common complaint regarding environmental legislation in Brazil, as elsewhere, is the existence of possible loopholes that open the door to abuse and differentiated responsibilities. Second-growth forests are perhaps particularly susceptible to this problem owing to their highly dynamic nature and the sensitivities of imposing clearance restrictions in places where some farmers rely on rotation-fallow systems in order to maintain their livelihoods.

An example of this potential problem is in the recent regulation, IN 02/2014, to determine the clearance of second-growth forest. The law states that areas younger than five years can be cleared irrespective of their physical structure, whilst areas older than 20 years must be conserved. Areas between five and 20 years can be licensed for clearance if the total basal area of the forest is less than 10 m^2ha^{-1}. However, in the absence of highly prescriptive guidance on how field surveys (to determine if the basal area of a site is above or below the threshold) should be conducted, landholders are able to position vegetation plots such that they avoid the largest and densest areas of trees. Moreover, if the area is subject to selective logging or allowed to burn prior to conducting the field surveys, it is possible that the average basal area could be reduced just beneath the critical threshold, thus deceiving the regulators. Indeed, there is some anecdotal evidence that this is already happening.

3.2. Operational Impediments to Achieving Good Governance of Second-Growth Forests

In addition to problems of clarity and consistency associated with the legal framework itself, there are significant operational impediments to the implementation of effective legislation governing existing and regenerating second-growth forests. Here, we consider some of the practical difficulties associated with restoration forestry regarding: (i) an understanding of the historical, geographical and ecological aspects that influence the likely success of any restoration project; (ii) access to technical and institutional support for restoration activities; (iii) the availability of and access to resources for monitoring; and (iv) access to adequate financial resources.

3.2.1. Understanding of Historical, Geographical and Ecological Aspects That Influence the Likely Success of any Restoration Project

As is the case across many parts of the tropics, there is limited knowledge on many of the ecological factors that could influence the success of restoration in different areas of a highly heterogeneous region, such as Pará. Some previous research has addressed how past land uses [24] and natural conditions influence forest regeneration, such as climate factors [25] and soil fertility [26]. However, further research is needed to help elucidate the positive or negative influence of factors, such as previous land-use intensity, the availability of nearby forests to act as a source for seeds, the use of different species of nurse trees and how to adapt restoration to suit particular ecological contexts, such as forests on white-sand soils, steep slopes and in riparian areas.

3.2.2. Access to Technical Guidance and Institutional Support for Restoration Activities

The success of a given restoration project is often determined by technical and institutional factors. Landowners commonly report that the lack of technical guidance to advise on planting and management techniques is more of a barrier to restoration than problems in accessing credit. This is particularly the case in designing restoration projects for areas where natural succession may be inhibited or to ensure that regenerating stands include economically valuable species, such as fruit and timber trees. There are major knowledge gaps regarding the species of seeds or seedlings that are likely to be established under different environmental conditions and levels of degradation. Although some federal institutions (Empresa Brasileira de Pesquisa Agropecuária and the Universidade Federal Rural do Pará) participate in the National Network of Seeds and have protocols for collecting the seeds of native forests species, the availability of these seeds remains limited by a lack of qualified collectors. Similarly, access to sufficient seedlings can often be a critical factor limiting the restoration of large areas with native trees. Problems related to technical assistance are often exacerbated by a lack of clarity in technical guidelines as to what is permissible, as well as inconsistencies in prescriptions given by different levels of government. An example here is the lack of clarity in what defines a native *versus* a non-native species, despite the fact that this distinction is often invoked in regulations governing forest restoration. Moreover, there are a number of provisions for using mixed plantings of native and non-native species, including in both the restoration of legal reserves, but also riparian habitat (where non-native species can be used temporarily to aid regeneration), without the necessary detail on the number and type of native species that should be used or the extent to which non-native trees are permissible. Another important technical limitation is the number of trained personnel who are able to conduct both desk and field assessments to appropriately map, sample and classify regenerating forests into different stages (*i.e.*, in accordance with the prescriptions given by IN 02/2014).

Of course, the full set of challenges and opportunities linked to restoration in developing countries go far beyond technical concerns related to the restoration process, but also relate fundamentally to the social dimensions of forest restoration, including opportunities for improving livelihoods and food security in rural communities [27,28].

3.2.3. Resources for Monitoring Second-Growth Forests

Although Brazil is one of the few countries to publish spatially explicit deforestation information every year, comparable information on second-growth forests is not yet available. In 2008, Brazil launched a new land-use monitoring project for the Amazon, called TerraClass, that provides biannual data on different production systems, early- and late-stage second-growth and primary forest. This is a useful contribution, but in order to guide fair and consistent decision-making regarding which areas of forest should be conserved and which can be licensed for clearance, a full time series analysis is needed to generate a map of the age of different second-growth stands for the whole state.

3.2.4. Access to Adequate Financial Resources

The revised Forest Code calls for the establishment and promotion of credit lines and extension services to support forest restoration work. Great expectations have been associated with the Programa de Regularização Ambiental (PRA, Table 1) in this regard, but as of yet, no specific incentives or support for restoration have been offered. Whilst access to credit for restoration has improved in the past five years, there is still relatively little awareness as to what opportunities exist and what criteria need to be satisfied in order to access this credit. After the revision of the Forest Code, new credit lines were launched, such as within the smallholder-dedicated Plano Safra dedicated to smallholders, which provided resources specifically designated for forest restoration in LRs and APPs. The barriers to accessing credit are often related to uncertainty and disputes over land titles, and the costs and technical resources needed to resolve tenure problems in the dynamic frontier landscapes that characterise much of Pará, as other areas of the tropics, can be enormous.

4. The Challenge of Governing Second-Growth Forests in the Tropics

As discussed at the start of this paper, second-growth forests present a particular governance challenge, both because they represent what are often highly dynamic components of complex mosaic landscapes, but also because their value for conservation and society is often poorly appreciated by many key actors. In many ways, second-growth forests epitomize the tensions that commonly exist between environmental and agricultural sectors. On the one hand, the fact that second-growth forests represent a critical component of fallow-based agricultural systems, including millions of traditional smallholder farmers throughout the world, means a strict "fences and fines" conservation agenda is not appropriate. Yet, on the other hand, the fact that these forests are critical to the provision of local and regional ecosystem services, including pollination, soil conservation and the maintenance of hydrological systems, as well as the protection of globally important biodiversity, demands that the restoration of degraded areas is made a major environmental policy, as indeed, it has been in Brazil under the revised Forest Code. Integrating these priorities with trajectories of agricultural development is particularly challenging in highly biodiverse frontier regions, such as Pará, that host a highly diverse array of actors, including millions of poor and vulnerable smallholder farmers.

In seeking to overcome this challenge, we have highlighted the importance of legal and operational impediments that are typical of secondary forest governance worldwide [27–29]. One overarching recommendation that emerges from our analysis and that is echoed by studies elsewhere is the need for much greater clarity, consistency and transparency in the rules that govern the conservation and restoration of second-growth forests, recognizing that the dynamic and uncertain nature of these forests makes this particularly challenging compared to other areas of environmental governance. We argue that achieving this is only possible through careful dialogue between researchers, policy makers and societal representatives involved in both the environmental and agricultural sectors. Second-growth forests, perhaps more than any other area of land management, require a landscape approach that places the costs and benefits of the full land use mosaic at the heart of decision-making. We also recognize the importance of actors and institutions capable of bridging sectors and levels of governance and ensuring that possible contradictions, inconsistencies and flaws are identified and resolved openly. Brazil and, in particular, the state of Pará have made significant progress towards this with the nation-wide consultation process that underpinned the revision of the Forest Code and the establishment of cross-sectoral agencies, such as Pará's Green County Program. Significant work still remains in Pará, as elsewhere, to ensure that this process gains sufficient momentum to establish a system that links forest conservation and agricultural agendas. This needs to be done by bringing together the regulations, incentives, technical support and monitoring instruments capable of fostering a lasting and fair approach to the management of second-growth forests for the benefit of future generations. Moreover, it is vital that second-growth forests are ultimately recognized as being a benefit, rather than impediment, to the development of sustainable agricultural and forestry systems, ensuring the maintenance of critical ecosystem services, the conservation of biodiversity and the provision of a large and poorly developed job market in restoration ecology and forest management [27–29]. With this in mind, the assessment of second-growth forest conservation and restoration programs should not be made based on technical indicators of forest condition alone, but should incorporate an understanding of the drivers of success, encompassing the suite of inter-related biophysical, socioeconomic and political-institutional factors that will ultimately determine the success or failure of a given project [27].

Acknowledgments

We are grateful to the Instituto Nacional de Ciência e Tecnologia, Biodiversidade e Uso da Terra na Amazônia (CNPq grant 574008/2008-0), for financial support; CNPq and CAPES for the Science without Borders Program's Fellowship to ACL and JB, as well as the CNPq for Research Productivity grant (CNPq 306368/2013-7) to ICGV and the CNPq grant 479429/2013-8 to JF; and to support from the Swedish Research Council Formas (Grant 2013-1571) to TG. We are grateful to Dr Marcos Adami and Marcia Barros, from INPE-Centro Regional da Amazônia, for preparing the map of secondary forests, and to Amir Sokolowski, for useful discussions about environmental legislation and the legislative process. This is paper number 26 from the Sustainable Amazon Network [30]. We greatly appreciate the useful comments of the reviewers and editors of the special issue on the earlier draft of this manuscript.

Author Contributions

All authors contributed to the planning, research and writing of this paper.

Conflicts of Interest

The authors declare no conflict of interest.

References

1. Wright, S.J. The future of tropical forests. *Ann. NY Acad. Sci.* **2010**, *1195*, 1–27.
2. Asner, G.P.; Rudel, T.K.; Aide, T.M.; Defries, R.; Emerson, R. A contemporary assessment of change in humid tropical forests. *Conserv. Biol.* **2009**, *23*, 1386–1395.
3. Barlow, J.; Gardner, T.A.; Araujo, I.S.; Ávila-Pires, T.C.; Bonaldo, A.B.; Costa, J.E.; Esposito, M.C.; Ferreira, L.V.; Hawes, J.; Hernandez, M.I.M.; *et al.* Quantifying the biodiversity value of tropical primary, secondary, and plantation forests. *Proc. Natl. Acad. Sci. U.S.A.* **2007**, *104*, 18555–18560.
4. Chazdon, R.L.; Peres, C.A.; Dent, D.; Sheil, D.; Lugo, A.E.; Lamb, D.; Stork, N.E.; Miller, S. Where are the wild things? Assessing the potential for species conservation in tropical second-growth forests. *Conserv. Biol.* **2009**, *23*, 1406–1417.
5. Moura, N.G.; Lees, A.C.; Andretti, C.B.; Davis, B.J.; Solar, R.R.; Aleixo, A.; Barlow, J.; Ferreira, J.; Gardner, T.A. Avian biodiversity in multiple-use landscapes of the Brazilian Amazon. *Biol. Conserv.* **2013**, *167*, 339–348.
6. Szott, L.T.; Palm, C.A.; Buresh, R.J. Ecosystem fertility and fallow function in the humid and subhumid tropics. *Agrofor. Syst.* **1999**, *47*, 163–196.
7. Feldpausch, T.R.; Rondon, M.A.; Fernandes, E.C.; Riha, S.J.; Wandelli, E. Carbon and nutrient accumulation in second-growth forests regenerating on pastures in central Amazonia. *Ecol. Appl.* **2004**, *14*, 164–176.
8. Suding, K.N. Toward an era of restoration in ecology: Successes, failures, and opportunities ahead. *Ann. Rev. Ecol. Syst.* **2011**, *42*, 465.
9. Asner, G.P.; Powell, G.V.; Mascaro, J.; Knapp, D.E.; Clark, J.K.; Jacobson, J.; Kennedy-Bowdoina, T.; Balajia, A.; Paez-Acosta, G.; Victoriac, E.; *et al.* High-resolution forest carbon stocks and emissions in the Amazon. *Proc. Natl. Acad. Sci. USA* **2010**, *107*, 16738–16742.
10. Brown, S.; Lugo, A. Tropical second-growth forests. *J. Trop. Ecol.* **1990**, *6*, 1–32.
11. Fearnside, P.M.; Guimarães, W.M. Carbon uptake by second-growth forests in Brazilian Amazonia. *For. Ecol. Manag.* **1996**, *80*, 35–46.

12. Berenguer, E.; Ferreira, J.; Gardner, T.A.; Aragão, L.E.O.C.; Camargo, P.B.; Cerri, C.E.; Durigan, M.; Oliveira, R.C.; Vieira, I.C.; Barlow, J. A large-scale field assessment of carbon stocks in human-modified tropical forests. *Glob. Change Biol.* **2014**, doi:10.1111/gcb.12627.

13. Van Breugel, M.; Hall, J.S.; Craven, D.; Bailon, M.; Hernandez, A.; Abbene, M.; van Breugel, P. Succession of Ephemeral Secondary Forests and Their Limited Role for the Conservation of Floristic Diversity in a Human-Modified Tropical Landscape. *PLoS One.* **2013**, *8*, e82433. doi:10.1371/journal.pone.0082433.

14. Gardner, T.A.; Barlow, J.; Chazdon, R.; Ewers, R.M.; Harvey, C.A.; Peres, C.A.; Sodhi, N.S. Prospects for tropical forest biodiversity in a human-modified world. *Ecol. Lett.* **2009**, *12*, 561–582.

15. Hecht, S. The new rurality: Globalization, peasants and the paradoxes of landscapes. *Land Use Policy* **2010**, *27*, 161–169.

16. Penteado, A.R. *Problemas de Colonização e de Uso da TerranaRegiãoBragantina do Estado do Pará*; Universidade Federal doPará: Belém, Brizail, 1967; p. 488.

17. INPE. InstitutoNacional de Pesquisas Espaciais. *ProjetoTerraclass—Mapeamento da vegetação secundária para a Amazônia Legal*; INPE-CRA: Belém, Brazil, 2010. Available online: http://www.inpe.br/cra/terraclass.php# (accessed on 1 January 2014).

18. Neef, T.; Lucas, R.M.; dos Santos, J.R.; Brondizio, E.S.; Freitas, C.C. Area and age of secondary forests in Brazilian Amazonia 1978–2002: An empirical estimate. *Ecosystems* **2006**, *9*, 609–623.

19. Perz, S.G.; Skole, D.I. Social determinants of second-growth forests in the Brazilian Amazon. *Soc. Sci. Res.* **2003**, *32*, 25–60.

20. Salomão, R.P.; Vieira, I.C.G.; Brienza Junior, S.; Amaral, D.D; Santana, A.C. Sistema Capoeira Classe: Umaproposta de sistema de classificação de estágiossucessionais de florestassecundárias para o estado do Pará. *Bol. Mus. Para. Emilio Goeldi Cienc. Nat.* **2012**, *7*, 297–317.

21. Soares-Filho, B.; Rajao, R.; Macedo, M.; Carneiro, A.; Costa, W.; Coe, M.; Rodrigues, H.; Alencar, A. Cracking Brazil's Forest Code. *Science* **2014**, *344*, 363–364.

22. Holl, K.D.; Aide, T.M. When and where to actively restore ecosystems? *Forest Ecol. Manag.* **2011**, *261*, 1588–1563.

23. Lees, A.C.; Vieira, I.C.G. Oil-palm concerns in Brazilian Amazon. *Nature* **2013**, *497*, 188.

24. Uhl, C.; Buschbacher, R.; Serrão, E.A.S. Abandoned pastures In Eastern Amazonia. I. Patterns of Plant Succession. *J. Ecol.* **1988**, *76*, 663–681.

25. Vasconcelos, S.S.; Zarin, D.J.; Araújo, M.M.; de Sounza Miranda, I. Aboveground net primary productivity in tropical forest regrowth increases following wetter dry-seasons. *For. Ecol. Manag.* **2012**, *276*, 82–87.

26. Moran, E.F.; Brondizio, E.; Tucker, J.M.; da Silva-Forsberg, M.C.; Mccracken, S.D.; Falesi, I. Effects of soil fertility and land use on forest succession in Amazonia. *For. Ecol. Manag.* **2000**, *139*, 93–108.

27. Dinh le, H.; Smith, C.; Herbohn, J.; Harrison, S. More than just trees: Assessing reforestation success in tropical developing countries. *J. Rural Stud.* **2012**, *28*, 5–19.

28. Carabias, J.; Arriaga, V.; Cervantes Gutierrez, V. Las políticas públicas en la restauración ambiental en México: Limitantes, avances, rezagos y retos. *B Soc. Bot. Méx.* **2007**, *80*, 85–100.

29. Sharp, A.; Nakagoshi, N. Rehabilitation of degraded forests in Thailand: Policy and practice. *Landsc. Ecol. Eng.* **2006**, *2*, 139–146.

30. Rede Amazônia Sustentável: Assessing Land-Use Sustainability in Brazilian Amazon. Available online: www.redeamazoniasustentavel.org (accessed on 23 January 2014).

Re-Greening Ethiopia: History, Challenges and Lessons

Mulugeta Lemenih and Habtemariam Kassa

Abstract: In Ethiopia, deforestation rates remain high and the gap between demand and domestic supply of forest products is expanding, even though government-initiated re-greening efforts began over a century ago. Today, over 3 million hectares (ha) of degraded forest land are under area exclosure; smallholder plantations cover 0.8 million ha; and state-owned industrial plantations stagnate at under 0.25 million ha. This review captures experiences related to re-greening practices in Ethiopia, specifically with regards to area exclosure and afforestation and reforestation, and distills lessons regarding processes, achievements and challenges. The findings show that farmers and non-governmental organizations (NGOs) are the main players, and that the private sector has so far played only a small role. The role of the government was mixed: supportive in some cases and hindering in others. The challenges of state- and NGO-led re-greening practices are: inadequate involvement of communities; poorly defined rehabilitation objectives; lack of management plans; unclear responsibilities and benefit-sharing arrangements; and poor silvicultural practices. The lessons include: a more active role for non-state actors in re-greening initiatives; more attention to market signals; devolution of management responsibility; clear definition of responsibilities and benefit-sharing arrangements; and better tenure security, which are all major factors to success.

Reprinted from *Forests*. Cite as: Lemenih, M.; Kassa, H. Re-Greening Ethiopia: History, Challenges and Lessons. *Forests* **2014**, *5*, 1896-1909.

1. Introduction

Large areas of the world's forests have been lost or degraded, and the problem continues unabated. According to the Food and Agriculture Organization of the United Nations (FAO), around 13 million hectares (ha) of forest were converted to other uses or lost through natural causes each year between 2000 and 2010 compared to 16 million ha per year in the 1990s [1] though marked variations are observed across regions. Due to natural expansion and plantations, the annual net forest loss remains at about 5.2 million ha. The overall effect of such a loss and widespread forest degradation is a decline in environmental goods and services, including climate stabilization and loss of biodiversity and reduction in human well-being in general [2]. The fragile state of most tropical forests and the implications of forest degradation and deforestation are widely acknowledged and have been subject of discussion for several decades. Though not adequate to reverse the trend at a global level, various measures are being taken, including restoration and rehabilitation of degraded forest lands. South East Asian countries, notably China and Vietnam, have made significant gains in tree planting initiatives and reduced forest losses. Asia and South America account for 91% of the 4.5 million ha of annually planted area globally. In Africa, plantations are expanding, but at a much lower rate. Between 1990 and 2000, the area of forest plantation in Africa increased by less than 5%, while in Asia it grew by about 20%—from 45 million ha to 60 million ha [3]. In these Asian countries, the state is moving away from forest protection towards creating an enabling environment for non-state actors to play the lead role in plantation forestry [4]. In Africa, studies about forest rehabilitation efforts are scant.

Recently, Ethiopia has begun taking measures to rehabilitate degraded forests and forest lands. Deforestation is severe and has a long history in Ethiopia, especially in the central and northern highlands where subsistence farming and settlements have been changing landscapes for millennia. Most of the remaining natural high forests of the country are found in the southwest, which was remote and inaccessible until recently. Up to the beginning of the 20th century, people and political capitals tackled scarcity of forest products, notably wood, by moving close to forested landscapes [5]. However, in the 1890s, an alternative approach involving re-greening through reforestation and afforestation (RA) was promoted by the emperor of Ethiopia, Menilik-II [5]. This marks the first formal re-greening [6] attempt by the government in the history of Ethiopia.

A number of other factors also justify the need for re-greening in Ethiopia. The country is home to more than 90 million people. Over 90% of the population's energy requirement is obtained mainly from biomass [7]. Unsustainable harvest from natural forests and woodlands has reduced the supply of woody biomass, further widening the gap between supply and demand. The low level of industrial wood supply from in-country production is compensated by a large volume of imports. For instance, in 2010/11 Ethiopian Fiscal Year (that begins on July 7 2010 and ends on July 6 2011), the import bill for wood products reached [8] Birr 1.8 Billion (US$ 115 million), creating an additional challenge for a country struggling to increase its foreign currency earnings.

There is a growing recognition that deforestation and forest degradation should be reduced. In its strategy document of December 2011, the Government of Ethiopia (GoE) identified the forestry sector as one of the pillars of the green economy that the country is planning to build by 2030 [7]. The government also set the following major targets for the forestry sector: afforestation on 2 million ha, reforestation on 1 million ha and improved management of 3 million ha of natural forests and woodlands. Through proper management of 5 million ha of forests and woodlands, Ethiopia hopes to achieve 50% of its total domestic greenhouse gas (GHG) emissions abatement potential by 2030 [7]. To this end, the country is engaged in various re-greening undertakings, and plans to scale-up good practices. However, little is known about the effectiveness and sustainability of these re-greening practices. The objective of this paper is to explore the different re-greening practices in the Ethiopian forestry sector in order to capture experiences and distill lessons for governing forest restoration [9]. A review of literature and official reports, as well as discussions with key informants and experts, constituted the major means of generating information used in the study. Although most of re-greening practices covered in the paper do not satisfy the definition of forest restoration, the lessons may guide and improve actual and future programs devoted to assist the recovery of native ecosystems in Ethiopia and elsewhere. While much remains unknown about how these re-greening practices could be modified to enhance restoration objectives, this paper proposes options to make forest systems more sustainable.

2. Drivers, Agents and Objectives of Re-Greening Practices

In Ethiopia, demand for wood is increasing owing to population and economic growth. However, domestic supply continues to decline due to deforestation and low level of investment in plantation forests. Consequently, the gap between supply and demand is expanding. This has been perceived for many years and led to government-initiated re-greening efforts by the end of the 19th century. The principal drivers are: the rising demand and dwindling supplies of forest products; and increased recognition by policy makers of the importance of expanding forest cover to increase the supply of forest products, conserve biodiversity and reduce the decline in forest-based ecosystem services.

Though one would expect the state to be the lead agent in re-greening, it is mainly non-state actors, notably non-governmental organizations (NGOs), and farming households that are playing the major role in Ethiopia. The state influences the actions of these agents through its institutions and legal framework. In some cases, the state's policies are supportive of re-greening undertakings, while in other cases they are obstructive, e.g., rules constraining transportation of wood products from selected indigenous trees. Re-greening practices driven by NGOs and bilateral and United Nations (UN) agencies primarily emphasize environmental rehabilitation, while farmers undertake re-greening activities largely for economic gains with little, if any, focus on ecological objectives.

Why do non-state actors emphasize environmental objectives in their re-greening projects? The extent and severity of land degradation in Ethiopia is unprecedented. Major land-cover changes resulting from improper practices are taking place on the rugged topography that characterizes most of the Ethiopian highlands, which have accelerated land degradation and soil erosion. This has left vast areas severely degraded, while the loss of fertile topsoil, estimated at 1 billion cubic meters (m^3) per year, significantly reduces agricultural productivity and continues to threaten food security at household and national levels [10]. An earlier estimate by FAO [10] put the degraded area on the highlands at 27 million ha, of which 14 million ha are very seriously eroded and 2 million ha of the seriously eroded lands have reached a point of no return. This large-scale land degradation and its impact on agricultural productivity are believed to have contributed to the catastrophic famines that hit Ethiopia following droughts in the 1970s and 1980s. Consequently, UN agencies, notably the World Food Programme (WFP), alongside environmental NGOs, led initiatives for soil and water conservation as well as for forest land rehabilitation. Some of these rehabilitation projects later became national programs run by the government with financial assistance from donors. An example is the MERET Project (Managing Environmental Resources to Enable Transitions to More Sustainable Livelihoods Project), a national land rehabilitation program of the government supported by WFP. This also set the stage for a national Sustainable Land Management Program led by the Ministry of Agriculture and supported by a coalition of donors.

3. Major Types of Re-Greening Practices

Though re-greening practices in Ethiopia are diverse, this paper grouped them into two broad categories: area exclosure and afforestation/reforestation. Area exclosure is the dominant type of re-greening practice promoted by NGOs, as well as by multilateral and bilateral donors, on degraded lands whereas afforestation/reforestation includes small-scale and industrial plantations. Recently, the government has also begun promoting area exclosure activities across the country. These activities seek primarily to rehabilitate degraded forest land and its biodiversity, and ensure a continued supply of forest products and services. On the other hand, afforestation/reforestation activities through small-scale plantations are re-greening practices initiated and run by farmers themselves. Small-scale plantations mainly on degraded lands have become important particularly since the mid-1990s, while industrial plantations are still project-based state initiatives. The government continues to encourage industrial and peri-urban plantations to meet national industrial, construction and fuel wood demands.

4. Attributes, Challenges and Achievements of the Major Re-Greening Practices

4.1. Area Exclosures

Area exclosure is one of the most widespread forms of re-greening in Ethiopia today. It involves protecting areas mainly through social fencing from any form of cultivation, cutting trees and shrubs, or grazing by livestock. This is meant to allow regeneration and foster natural ecological succession for the rehabilitation of deforested areas or degraded forests. Two types of area exclosure management are observed. The first one involves no additional management activities other than protecting enclosed areas against livestock and human interference. Ecological succession will occur from buried or dispersed seeds. The second type, which is the most common, involves planting of seedlings (exotic or indigenous species), aerial seeding and construction of soil water conservation structures to speed up succession through the modification of microclimatic and soil conditions. Besides producing wood for subsistence and markets, planted trees create an environment conducive for nursing some indigenous tree species. As a result, diverse woody and non-woody plant species re-emerge, landscape greenness increases, soil erosion declines, sediment deposition downstream declines and water infiltration and stream discharge increase [11–15]. As the exclosure age increases, the density of woody species rises and canopy cover expands, suppressing herbaceous plants. Farmers find this discouraging since it significantly reduces the volume and quality of livestock feed harvested from area enclosures.

Re-greening through area exclosure is employed in a wide range of forest ecosystems—from dry forests and woodlands to the sub-humid Afromontane forests. In 1996, there were only about 143,000 ha of exclosure in Ethiopia [16]. However, in Tigray regional state alone the area under area exclosure reached 895,220 ha in 2011 [17]. Regional states are rapidly increasing areas put under exclosures, and by the end of 2013, exclosures covered 1.54 million ha in Tigray [18] and 1.55 m ha in Amhara [19].

4.2. Governance of Area Exclosures

Most area exclosures were not initiated either by the state or by communities, but were rather driven mainly by aid agencies and NGOs. At the beginning, there was little or no involvement of communities in defining the objectives and the course of the process. Prominent actors in this regard have been the WFP, the German aid agency GIZ (*Deutsche Gesellschaft fuer Internationale Zusammenarbeit*) and, more recently, NGOs associated with the ruling party like the Relief and Emergency Society of Tigray and the Organization for the Rehabilitation and Development of Amhara.

The soil and water conservation works associated with most area exclosure initiatives require huge labor investment, and need support to cover at least part of this cost. If not, these large-scale re-greening attempts are not feasible [16]. Over the last three years, the government has begun mobilizing communities to secure free labour during the dry season to undertake massive soil and water conservation work in their respective watersheds, and to plant rehabilitated areas and open spots in the watersheds with tree seedlings. Official reports indicate that annually large areas of land are rehabilitated and planted with hundreds of millions of seedlings. The value of farmers' labour in rehabilitating degraded lands and planting them with tree seedlings during the dry season of 2013/14 Ethiopian Fiscal Year, was estimated by the Ministry of Agriculture to be Birr 10 billion (~0.5 billion USD).

The management of area exclosures remains largely top-down. During the socialist *Derge* regime (1974–1991) government agencies made decisions, and communities were simply informed and expected to collaborate. Although the approach has changed slightly since 1991, proposals for area exclosures still come usually from government agencies or NGOs, and communities are consulted with the expectation they will agree to such proposals. Sites are supposed to be selected jointly, by involving communities, but development agents (DAs) of the District Office of Agriculture together with *kebelle* (the lowest administrative unit in Ethiopia) administrators reportedly dominate the process. Consulting with farmers implies convincing them to implement the development programs planned by the government in federal or regional capitals [20,21]. This is partly because government sets targets (e.g., areas under area exclosure) that the district and *kebelle* administrators have to meet. Thus, local authorities tend to push farmers to participate in such initiatives and to fulfill quotas imposed centrally upon them. Under certain circumstances, local authorities use strategies to ensure that households participate in these undertakings. Some use their administrative power on those who fail to participate to reduce their benefits from government support programs or to limit their access to credit and agricultural inputs [22]. This undoubtedly affects the outcome and sustainability of area exclosures.

Some exclosures are protected by guards paid by contributions made by communities managing area exclosures [22]. In addition, check points are established along roads to discourage transportation of wood from exclosures. Communities are allowed to use grass through cut and carry, and to harvest honey from bee hives placed inside area exclosures. Nearly all area exclosures lack management plans, and little work has been done to find out options to improve management of exclosures to speed up their re-greening process and also their economic returns. Hence, one would expect limited annual growth and low yield levels. Also, systematic studies are lacking to determine whether current incentives alone would outweigh the cost of establishing and managing exclosures and meet the expectations of communities involved.

Some believe that allowing communities to use exclosures will simply destroy them, and rehabilitating the environment needs to be seen as separate from people [22]. This does not seem to be the view of regional states [23]. For instance, in Tigray some 63,000 ha of hillside areas under exclosure have been taken from the community and allocated to landless youth to manage and use [18]. This, however, has been identified as a possible disincentive for engaging communities in area exclosure activities in the future. On the other hand, the Amhara Regional Bureau of Agriculture reported having transferred the management and ownership rights of 27,800 ha of state-owned plantation forests on degraded lands to the communities that were involved in their establishment and protection [19].

Based on experiences in area exclosure management, the following points are suggested to enhance the effectiveness of practices:

i. Policy makers and practitioners need to move from a purely environmental orientation towards also ensuring socio-economic benefits, since unmet community expectations are likely to be major challenges for sustaining area exclosures [24,25]. In some areas, communities complain that closing off area exclosures for many years is affecting their livelihoods negatively [22].

ii. All exclosures need to have negotiated and clearly defined objectives, as well as agreed-upon management plans. Currently, neither communities nor the government agencies know for how long the areas will remain closed, how they will be managed for better economic and ecological outcomes, and what indicators should be used to measure socio-economic and environmental gains.

iii. Devolving responsibilities to lower levels of community organizations are likely to result in better area exclosure management. Area exclosures are managed at various levels of community organization, ranging from individuals, to village, *kebelle* or district levels. These different levels vary in their degree of effectiveness to facilitate collective action. Gebremedhin *et al.* [24] reported that collective action was stronger and socio-economic benefits greater, among smaller groups such as villages than among higher-level ones such as districts. Consequently, most communities prefer to divide communal lands into smaller individual plots for better management, including tree planting [25,26]. This indicates that cooperation for re-greening practices is likely to be more effective among small groups as the members tend to be more homogenous.

iv. The management and user rights aspects of area exclosures need to be defined better and formal agreements made between government agencies and communities regarding their respective rights and responsibilities. Increased conflicts are reported between members of communities regarding access to and use of area exclosures and in recognizing and protecting their boundaries [17]. Recently, some regional states began allocating area exclosures to landless youth. This is likely to cause disappointment among communities that established and managed these area exclosures. This will also discourage communities from participating in such re-greening practices in the future.

v. Dependency on external support needs to be reduced. The activities of external organizations assisting the establishment and management of area exclosures may, under certain circumstances, reduce local effort to engage in re-greening activities as communities may expect external support to initiate and sustain collective actions in establishing and managing area exclosures.

4.3. Afforestation and Reforestation

The total area of plantation forests in Ethiopia is estimated at 972,000 ha (Table 1). Afforestation/reforestation (AR) practices are meant primarily to increase the supply of wood products in the country. These practices comprise mainly three forms: industrial plantation, peri-urban energy forestry and small-scale plantations (Table 1). The former two are mainly government-driven, while the third is undertaken principally by farming households.

Table 1. Area under Plantation Forests (ha) in Four Major Regional States of Ethiopia (Source: [27]).

Regional State	Industrial Plantations	Non-industrial Small-scale Private Plantations	Peri-urban Energy Plantations	Total
Oromia	78,800	27,800	26,700	133,300
Amhara	44,600	639,400		684,000
Southern Nations, Nationalities and Peoples	27,300	64,000		91,300
Tigray	39,700	23,700		63,400
Total	190,400	754,900	26,700	972,000

Major industrial plantations are found in south-central and south-western regions, while peri-urban plantations were established around major cities such as Addis Ababa, Adama, Dessie, Gondar and Bahir Dar, with support obtained from the United Nations Development Programme (UNDP), the Swedish International Development Agency (Sida) and the World Bank. In some cases, industrial plantations were established on degraded forest lands bordering remnant natural forests such as Munessa Shashamane and Belete Gera forests. These plantations had dual objectives of providing round industrial wood and reducing pressure on natural forests.

Small-scale plantations have expanded, especially since the 1970s when the number of farming households planting trees began increasing significantly [25,26]. These plantations cover an estimated area of 754,900 ha (Table 1), making the rural landscape greener than it was some decades ago. They supply the largest volume of wood products used in the construction sector (such as poles and posts) and a significant portion of the biomass fuel consumed in the country. Small-scale plantations are established for two purposes: to satisfy household demands for wood and to generate additional household income from sales. For instance, in the Arsi highlands of central Ethiopia, wood from *Eucalyptus* grown by smallholder farmers contributes to 92% of the poles, 74% of the timber, 85% of the firewood, 40% of the charcoal, 83% of the posts and 91% of the farm implements used by a rural household. It also accounted for 74% of firewood, 100% of poles, 100% of posts and 21% of charcoal coming to Huruta town on market days from surrounding rural areas [28]. Income from *Eucalyptus* sales contributes on average up to 25% of total household annual cash income [28–30], and for poor households up to 72% of the total annual cash income [28], which is the largest non-agricultural source of household income [31,32]. In some areas, high rate of return from plantations compared to other farm enterprises [26,31] is leading to the conversion of croplands and grazing fields to *Eucalyptus* woodlots [33]. Also, having a woodlot of *Eucalyptus*, bamboo and other tree species accords a household head good societal respect, better self-esteem and pride. For instance, among the Gurage community in southern Ethiopia having a *Eucalyptus* woodlot bestows a considerable reputation and social value to the owner, and this reputation grows as the size of the woodlot increases [34]. According to Gemechu [35], some farmers are also inspired to have woodlots by observing others who planted it, and to secure societal respect besides economic returns.

A limited number of species from four genera (*Eucalyptus, Cuppressus, Pinus* and *Acacia*) account for the majority of plantation forests in Ethiopia. *Eucalyptus*, in particular, covers more than 90% of the total planted forest area in Ethiopia [27]. Typical biological attributes that attract farmers to *Eucalyptus* include fast growth, coppicing ability, ease of management (such as non-palatability to cattle), established market demand for its wood, its ability to grow well even on degraded landscapes and its better growth performance than most indigenous tree species on degraded lands [31]. Moreover, farmers with limited farm sizes plant *Eucalyptus* in high density—up to 40,000 stems per hectare [28,33]—and yet stands show relatively good growth performance. *Eucalyptus* is also the first exotic tree species to be formally introduced to Ethiopia [5]. Since its introduction in the 1890s, its area coverage has expanded from about 5000 ha [36] to 894,240 ha in 2011 [27]. Close to 60 different species of the genus are reported to have been introduced to Ethiopia, but *E. globulus* and *E. camaldulensis* are the most widespread of all.

The dominance of exotic species in plantation development is also related to legal constraints. Policies to manage natural forests are mainly conservation-oriented. Thus, local communities are not allowed to use wood from natural forests for commercial purposes. They can use non-timber forest products (NTFPs) only. Harvesting and transporting of woods from some indigenous trees, including high-value indigenous timber tree species such as *Cordia africana*, are prohibited. Proponents of this restriction argue that in the absence of guidelines or a certification scheme to help distinguish between timber harvested from natural forests and that produced from trees on-farm, lifting restrictions would simply aggravate deforestation of natural forests. On the other hand, having indigenous species in plantations remains a disadvantage as farmers are prevented from harvesting and selling native timber. This is the case even though farmers have the knowledge to raise seedlings, plant and tend them, and that the economic value of these species in the long term could outweigh that of exotics. Ownership of indigenous trees outside forests, such as those on communal grazing lands, is also vaguely defined. Such policies discourage farmers from growing native timber species on their farm lands, and force them to continue planting mainly exotic species.

In contrast, recent changes in legal frameworks are having strong positive effects on re-greening. Prominent among these legal instruments is the Rural Land Administration Proclamation that improved tenure security among farming households through agricultural land registration and certification. As tenure insecurity was among the major deterrents for tree planting in the past [37], land certification improved the sense of tenure security, which in turn led to more tree planting [38–40]. Holden *et al.* [40], for instance, report a positive relationship between land certification and investment in land management, including tree planting. Similarly, Gebreegziabher *et al.* [38] note that tenure security is one of the major factors that positively and significantly affects tree planting and the amount of trees planted by rural households in Ethiopia. The forest policy, issued in 2007, also encourages tree planting as it proposes tax incentives to farmers for planting trees. Fearing that plantations will expand and take over productive agricultural fields, some regional states discourage farmers from planting *Eucalyptus*. These measures are not popular with farmers, and researchers are challenging their rationale. Recently, however, emergence of insect pests affecting *Eucalyptus* seedlings [41] has raised the concern of authorities and farmers. Thus, the challenges associated with the expansion of mono species plantations need to be identified and addressed.

5. Conclusions and Implications

5.1. Conclusions

Assessment of the major re-greening practices and their performance in Ethiopia reveal the following:

i. Ensuring that re-greening practices generate sufficient economic incentives for communities is key to their sustainability. If re-greening is only for environmental goals, it is less likely to encourage the participation of communities, especially the poor. Poor households can hardly afford to lose short-term economic gains for long-term environmental benefits unless they are properly compensated for that loss. When individuals are likely to generate direct and tangible benefits, they will be motivated to participate in re-greening initiatives, be it individually or collectively. Also, community participation needs to be inclusive and equitable.

ii. Getting the policy environment right is crucial. The commitment of the Government of Ethiopia to rehabilitate degraded lands is indicated in recent documents such as the Climate-Resilient Green Economy Strategy, the Rural Land Administration Proclamation and the current five-year development plan. Policies like land certification are having a strong positive effect on re-greening practices. Better tenure security, clear user rights, and devolution of responsibilities to lower levels of organization (individual household or smaller community) help facilitate collective action for better re-greening initiatives in communal systems.

iii. Market signals play important roles. Markets have been the major driving force behind the expansion of small-scale plantations across the highlands of Ethiopia. High return on investment in plantations is driving the conversions of even farm and grazing lands to woodlots in some areas in the central and western highlands. In some cases, however, markets—especially the labor market—may negatively influence plantations by increasing the opportunity cost of labor.

iv. The role of non-state actors was important in re-greening Ethiopia. The non-state actors, notably NGOs, played a key role in initiating and supporting re-greening practices, notably area exclosures. NGOs also advocated for policy reforms. However, since they were hardly learning from each other, some contradictory messages were given to communities and policy makers (e.g., on harvesting wood from natural forests). This undermined their capacity to help policy makers make informed decisions.

5.2. Implications for Policy and Practice

Based on the findings, the following measures are suggested to improve policy and practice to enhance the effectiveness of re-greening practices:

i. Policy makers need to consider the likely impacts of allocating communally-managed area exclosures to individuals. Recently, some regional governments have started to allocate to landless youths area exclosures that have been rehabilitated and managed by local communities. This could erode the trust of the community in the government in terms of ownership arrangement of area exclosures, and may affect their willingness to be engaged in re-greening communally-owned but degraded forest lands in the future. Thus, we recommend closer examination of the impacts of such exercises.

ii. The influence of the current policy and the prevailing market signals on re-greening practices and on the sustainability of impacts needs to be investigated further. Policies that encourage farmers to plant indigenous tree species and enable them to sell native woods from sustainably managed plantations or restored forests are needed, along with a mechanism of control so that such policies do not aggravate degradation and deforestation of natural forests.

iii. More incentives should be put in place, especially in terms of access to land and credit, to encourage private sector engagement in re-greening practices. Value-added processing options to increase returns from re-greening practices need to be explored and supported.

iv. Linking research to policy should be given special attention and knowledge needs to be translated into practice to enhance effectiveness, efficiency and equity aspects of re-greening practices.

v. Re-greening practices in Ethiopia lack coordination, both technically and managerially. Capacity building on restoration research in general, and on nationally important re-greening practices in particular, is critically needed. This would enable measures to enhance economic returns to communities and help identify, test and promote options to better achieve restoration objectives. Studies so far are exploratory and simply describe current vegetation succession, and how to speed it up. Also clearly lacking are indicators of success, and key silvicultural practices for management plans. Such practices could guide operations to enhance succession of vegetation types, as well as ensure benefits to local people and the environment. Re-greening practices must provide optimum ecological and socio-economic benefits. Achieving these objectives by managing trade-offs requires concerted efforts of researchers, development practitioners and policy makers.

Acknowledgments

The authors gratefully acknowledge the editors and reviewers for their comments and suggestions that helped improve the quality of the paper. We also thank CGIAR Forests Trees and Agroforestry research program for covering part of the costs associated with the work that led to this publication.

141

Author Contributions

Both authors contributed equally in writing the paper and approved the accepted version.

Conflicts of Interest

The authors declare no conflict of interest.

References and Notes

1. FAO (Food and Agriculture Organization of the United Nations). The global forest resources assessment 2010 main report (FRA 2010). *For. Pap.* **2010**, *163*, 13–31.
2. Lamb, D.; Gilmour, D. *Rehabilitation and Restoration of Degraded Forests*; IUCN and WWF: Gland, Switzerland, 2013.
3. Food and Agriculture Organization of the United Nations (FAO). Forestry outlook study for Africa—Regional report: Opportunities and challenges towards 2020. *For. Pap.* **2003**, *141*, 1–39.
4. Nawir, A.A.; Kassa, H.; Sandewall, M.; Dore, D.; Campbell, B.; Ohlsson, B.; Bekele, M. Stimulating smallholder tree planting—Lessons from Africa and Asia. *Unasylva* **2007**, *58*, 53–59.
5. Bekele, M. Forest Property Rights, the Role of the State, and Institutional Exigency: The Ethiopian Experience. Ph.D. Thesis, Swedish University of Agricultural Sciences, Uppsala, Sweden, 2003.
6. Re-greening is used in this paper to encompass all forms of management practices and other measures intended to rehabilitate degraded forestlands or to develop new forests through afforestation and reforestation.
7. FDRE (Federal Democratic Republic of Ethiopia). *Ethiopia's Climate-Resilient Green Economy Strategy*; Federal Democratic Republic of Ethiopia: Addis Ababa, Ethiopia, 2011; pp. 4–44.
8. Birr is the local currency in Ethiopia, and currently one Birr is exchanged for 0.051 USD.
9. Forest restoration describes activities to assist recovery of native ecosystems, and is defined by the Society of Ecological Restoration as a process of re-establishing the presumed structure, productivity and species diversity of the forest originally present at a site so as to closely match the ecological processes and functions of the restored forest to those of the original forest.
10. Food and Agricultural Organization of the United Nations (FAO). *Ethiopian Highland Reclamation Study* (*EHRS*); Final Report; FAO: Rome, Italy, 1984; pp. 1–154.
11. Desscheemaeker, K.; Nyssen, J.; Rossi, J.; Poesen, J.; Haile, M.; Raes, D.; Muys, B.; Moeyersons, J.; Deckers, S. Sediment deposition and pedogenesis in exclosures in Tigray Highlands, Ethiopia. *GEODERMA* **2006**, *132*, 291–314.
12. Yami, M.; Gebrehiwot, K.; Moe, S.; Mekuria, W. Impact of area enclosures on density, diversity, and population structure of woody species: The case of May Ba'ati-Douga Tembien, Tigray, Ethiopia. *Ethiop. J. Nat. Res.* **2006**, *8*, 99–121.

13. Yami, M.; Gebrehiwot, K.; Moe, S.; Mekuria, W. Impact of area enclosures on density and diversity of large wild mammals: The case of May Ba'ati, Douga Tembien district, Central Tigray, Ethiopia. *East Afr. J. Sci.* **2007**, *1*, 55–68.

14. Babulo, B.; Muys, B.; Nega, F.; Tollens, E.; Nyssen, J.; Deckers, J.; Mathijs, E. Household livelihood strategies and forest dependence in the highlands of Tigray, Northern Ethiopia. *Agric. Syst.* **2008**, *98*, 147–155.

15. Mekuria, W.; Veldkamp, E.; Tilahun, M.; Roland, O. Economic valuation of land restoration: The case of exclosures established on communal grazing lands in Tigray, Ethiopia. *Land Degrad. Dev.* **2011**, *22*, 334–344.

16. EFAP (Ethiopian Forestry Action Program). *Ethiopian Forestry Action Program*; EFAP: Addis Ababa, Ethiopia, 1993; pp. 10–25.

17. Yami, M.; Mekuria, W.; Hauser, M. The effectiveness of village bylaws in sustainable management of community-managed exclosures in Northern Ethiopia. *Sustain. Sci.* **2012**, *8*, 73–86. Available online: http://link.springer.com/article/10.1007/s11625-012-0176-2 (accessed on 5 April 2014).

18. Gebresilassie, M. Forest development endeavors in Tigray region: Opportunities, challenges and way forward. In Proceedings of the Workshop Organized by CIFOR Ethiopia Office, Addis Ababa, Ethiopia, 29 December 2013.

19. Bureau of Agriculture (BoA). *Forest Development and Utilization in Amhara National Regional State: Issues Requiring Attention*; Presented to a workshop organized by CIFOR Ethiopia Office, Addis Ababa, Ethiopia, 29 December 2013.

20. Segers, K.; Dessein, J.; Nyssen, J.; Haile, M.; Deckers, J. Developers and farmers intertwining interventions: The case of rainwater harvesting and food-for-work in Degua Temben, Tigray, Ethiopia. *Int. J. Agric. Sustain.* **2008**, *6*, 173–182. Available online: http://www.tandfonline.com/doi/abs/10.3763/ijas.2008.0366#.U6wq5v36jIU (accessed on 6 March 2014).

21. Segers, K.; Dessein, J.; Hagberg, S.; Develtere, P.; Haile, M.; Deckers, J. Be like bees—The politics of mobilizing farmers for development in Tigray, Ethiopia. *Afr. Aff.* **2009**, *108*, 91–109. Available online: http://afraf.oxfordjournals.org/content/108/430/91.short (accessed on 12 December 2013).

22. Gebremichael, Y.; Waters-Bayers, A. *Trees are Our Backbone: Integrated Environment and Local Development in Tigray Region of Ethiopia*; International Institute for Environment and Development (IIED): London, UK, 2007; pp. 1–24.

23. Regional state refers to constituent administrative structure of the Federal Democratic Republic of Ethiopia. The nine regional states, which have their own executive (government) and legislative authorities (parliament), have considerable decision-making power in budget allocations and management of natural resources including land and forest.

24. Gebremedhin, B.; Pender, J.; Tesfay, G. Community natural resource management: The case of woodlots in Northern Ethiopia. *Environ. Dev. Econ.* **2003**, *8*, 129–148. Available online: http://journals.cambridge.org/action/displayAbstract? (accessed on 2 November 2013).

25. Wisborg, P.; Shylendra, H.S.; Gebrehiwot, K.; Shanker, R.; Tilahun, Y.; Nagothu, U.S.; Tewoldeberhan, S.; Bose, P. Rehabilitation of common property resources through re-crafting of village institutions: A comparative study from Ethiopia and India. In Proceedings of the Eighth Biennial Conference of the International Association for the Study of Common Property (IASCP), Bloomington, Indiana, 31 May–4 June 2000; Centre for International Environment and Development Studies (Noragric), Agriculture University of Norway: Oslo, Norway, 2000.

26. Lemenih, M. Growing eucalypts by smallholder farmers in Ethiopia. In Proceedings of the Conference on Eucalyptus Species Management, History, Status and Trends in Ethiopia, Addis Ababa, Ethiopia, 15–17 September 2010; Gil, L., Tadesse, W., Tolosana, E., Lopez, R., Eds.; Ethiopian Institute of Agricultural Research (EIAR): Addis Ababa, Ethiopia, 2010; pp. 91–103.

27. Bekele, M. Forest plantations and woodlots in Ethiopia. *Afr. For. Forum Work. Pap. Ser.* **2011**, *1*, 1–51.

28. Mekonnen, Z.; Kassa, H.; Lemenih, M.; Campbell, B. The role and management of eucalyptus in Lode Hetosa district, Central Ethiopia. *For. Trees Livelihoods* **2007**, *17*, 309–323.

29. Tesfay, T. Problems and Prospects of Tree Growing by Smallholder Farmers. A Case Study in Feleghe-Hiwot locality, Eastern Tigray. Master's Thesis, Swedish University of Agricultural Sciences, Skinnskatteberg, Sweden, 1996.

30. Minda, T. Economics of Growing *Eucalyptus Globulus* on Farmers' Woodlots: The Case of Kutaber District, South Wollo, Ethiopia. Master's Thesis, Hawassa University, Wondo Genet College of Forestry, Awassa, Ethiopia, 2004.

31. Jagger, P.; Pender, J. The role of trees for sustainable management of less-favored lands: The case of *Eucalyptus* in Ethiopia. *For. Pol. Econ.* **2003**, *5*, 83–95. Available online: http://www.sciencedirect.com/science/article/pii/S1389934101000788 (accessed on 15 January 2014).

32. Turnbull, J.W. *Eucalyptus* plantations. *New For.* **1999**, *17*, 37–52.

33. Jenbere, D.; Lemenih, M.; Kassa, H. Expansion of *Eucalypt* farm forestry and its determinants in Arsi Negelle District, South Central Ethiopia. *Small-Scale For.* **2011**, *11*, 389–405. Available online: http://link.springer.com/article/10.1007/s11842-011-9191-x (accessed on 23 October 2013).

34. Achalu, N. Farm forestry decision making strategies of the Guraghe households, southern-central highlands of Ethiopia. Ph.D. Thesis, Institut fur Internationale Forst-Und Holzwirtschaft Technische Universitat, Dresden, Germany, 2004.

35. Gemechu, T. Expansion of *Eucalyptus* plantations by smallholder farmers amid natural forest depletion: Case study from Mulo District in Central Oromia, Wondo Genet College of Forestry and Natural Resources. In Proceedings of the Conference on Eucalyptus Species Management, History, Status and Trends in Ethiopia, Addis Ababa, Ethiopia, 15–17 September 2010; Gil, L., Tadesse, W., Tolosana, E., Lopez, R., Eds.; Ethiopian Institute of Agricultural Research (EIAR): Addis Ababa, Ethiopia; pp. 335–350.

36. Getahun, A. *Eucalyptus* farming in Ethiopia: The case for *Eucalyptus* woodlots in the Amhara Region. In Proceedings of the Conference on Eucalyptus Species Management, History, Status and Trends in Ethiopia, Addis Ababa, Ethiopia, 15–17 September 2010; Gil, L., Tadesse, W., Tolosana, E., Lopez, R., Eds.; Ethiopian Institute of Agricultural Research (EIAR): Addis Ababa, Ethiopia; pp. 206–221.

37. Kassa, H.; Bekele, M.; Campbell, B. Reading the landscape past: Explaining the lack of on-farm tree planting in Ethiopia. *Environ. Hist.* **2011**, *17*, 461–479.

38. Gebreegziabher, Z.; Mekonnen, A.; Kassie, M.; Kohlin, G. *Household Tree Planting in Tigrai, Northern Ethiopia: Tree Species, Purposes, and Determinants*; Environment for Development, Discussion Paper Series EfD DP 10-01; Ethiopian Development Research Institute: Addis Ababa, Ethiopia, January 2010; pp. 1–29.

39. Mekonnen, A. Tenure security, resource endowments, and tree growing: Evidence from the Amhara region of Ethiopia. *Land Econ.* **2009**, *85*, 292–307.

40. Holden, S. From Being Property of Men to Becoming Equal Owners? Early Impacts of Land Registration and Certification on Women in Southern Ethiopia; Final Research Report Prepared for UN HABITAT Shelter Branch; Land Tenure and Property Administration Section: Nairobi, Kenya, 2 January 2008; pp. 11–79.

41. Gezahgne, A.; Cortinas, M.-N.; Wingfield, M.J.; Roux, J. Characterisation of the Coniothyrium stem canker pathogen on *Eucalyptus camaldulensis* in Ethiopia. *Australas. Plant Pathol.* **2005**, *34*, 85–90.

Governing and Delivering a Biome-Wide Restoration Initiative: The Case of Atlantic Forest Restoration Pact in Brazil

Severino R. Pinto, Felipe Melo, Marcelo Tabarelli, Aurélio Padovesi, Carlos A. Mesquita, Carlos Alberto de Mattos Scaramuzza, Pedro Castro, Helena Carrascosa, Miguel Calmon, Ricardo Rodrigues, Ricardo Gomes César and Pedro H. S. Brancalion

Abstract: In many human-modified tropical landscapes, biodiversity conservation and the provision of ecosystem services require large-scale restoration initiatives. Such initiatives must be able to augment the amount and the quality of remaining natural habitats. There is thus a growing need for long-term, multi-stakeholder and multi-purpose initiatives that result in multiple ecological and socioeconomic benefits at the biome scale. The Atlantic Forest Restoration Pact (AFRP) is a coalition of 260+ stakeholders, including governmental agencies, private sector, NGOs and research institutions, aimed at restoring 15 million ha of degraded and deforested lands by 2050. By articulating, and then integrating common interests, this initiative has allowed different sectors of society to implement an ambitious vision and create a forum for public and private concerns regarding forest restoration. The AFRP adopts a set of governance tools so multiple actors can implement key processes to achieve long-term and visionary restoration goals. Having overcome some initial challenges, AFRP now has to incorporate underrepresented stakeholders and enhance its efforts to make forest restoration more economically viable, including cases where restoration could be less expensive and profitable. The AFRP experience has resulted in many lessons learned, which can be shared to foster similar initiatives across tropical regions.

Reprinted from *Forests*. Cite as: Pinto, S.R.; Melo, F.; Tabarelli, M.; Padovesi, A.; Mesquita, C.A.; de Mattos Scaramuzza, C.A.; Castro, P.; Carrascosa, H.; Calmon, M.; Rodrigues, R; *et al.* Governing and Delivering a Biome-Wide Restoration Initiative: The Case of Atlantic Forest Restoration Pact in Brazil. *Forests* **2014**, *5*, 2212-2229.

1. Introduction

In many human-modified tropical landscapes, the conservation of biodiversity and the provision of ecosystem services require innovative, large-scale restoration initiatives, which should seek to augment the amount/quality of natural habitats via the inclusion of both remaining forest patches and those undergoing restoration [1]. However, governments have only recently started to develop environmental policies aimed at reducing deforestation and promoting reforestation. Many countries are addressing their environmental problems and, more recently, their need to increase native vegetation cover through state-led and complex legal/regulatory instruments, which could (a) be excessively bureaucratic, (b) operate via top-down approaches, and (c) focus on legal compliance and punishment, instead of rewarding positive actions. Such approaches have failed to encourage better practices, resulting in low involvement and a lack of participation among multiple stakeholders, especially in regions with poor governance and weak legal enforcement [2]. In the context of ecological restoration initiatives in developing countries, a bottom-up approach could create good opportunities to overcome some of the legal, technological, and economic challenges frequently experienced by these initiatives [3,4]. In this context, the ambitious goal established by the Aichi Target 15 of the United Nations Convention on Biological Diversity to restore 15% of all degraded ecosystems on Earth by 2020 (about 150 million ha), as well as by the Bonn Challenge, requires well-coordinated and articulated initiatives [5]. As only a few countries, such as South Africa, the United States of America, Ethiopia, China and Costa Rica [6–8], have already started to implement large-scale initiatives, little information is available concerning instruments of governance and the coordination of restoration initiatives. It is imperative, therefore, that any lessons learned through both successful and unsuccessful experiences should be shared for the sake of large-scale forest restoration initiatives worldwide [9].

We describe here the socio-ecological context, the instruments of governance and the key challenges/lessons experienced by the Atlantic Forest Restoration Pact (hereafter AFRP), a biome-wide restoration program that represents the largest forest restoration initiative currently being implemented in Latin America [10]. We first address the degradation of the Brazilian Atlantic Forest and offer a historical perspective on the legal instruments and policies related to ecological restoration in this irreplaceable biome. We then contextualize why and how the AFRP was created, and discuss the governance structure specifically designed to achieve the AFRP's major goals and objectives in a dynamic environment of both opportunities and potential constraints. Finally, we highlight the major achievements of this restoration initiative and share the present and future challenges towards the implementation of this large-scale, multi-stakeholder forest restoration program with a view to inspiring and fostering similar initiatives across other tropical regions.

2. The History of Atlantic Forest Degradation

Even before the Portuguese settlers arrived in Brazil in 1500, the Atlantic Forest was already subject to some level of anthropogenic disturbance. The biome had become quite densely populated during the apex of the *Tupi* domination—a heterogeneous indigenous group that dominated the Brazilian Atlantic coast for approximately 1000 years before the arrival of European settlers—reaching around 600 people per 70 km² [11]. The *Tupi* people practiced nomadic slash-and-burn agriculture, and may have burned their entire territory—which was in the Atlantic forest biome—every 55 years (*i.e.,* during 1000 years of *tupi* domination, each forest patch appropriate for agriculture was probably burned dozens of times [11]). However, the site-specific and sporadic nature of this cultivation system did not impact the Atlantic Forest significantly and allowed for its vigorous re-growth after the *Tupi* societies collapsed.

Once the European settlers did not immediately find gold and silver to provide income to the Portuguese crown, they overexploited Brazilwood trees (*Caesalpinia echinata*) as a source of red dye for cloth, impacting nearly 600,000 ha of forest in the first century of European occupation [11]. The country's name derives from this endemic tree species of the Atlantic Forest, a species currently threatened by extinction. Concomitantly, the Portuguese crown provided land concessions in order to encourage people to consolidate the occupation of Brazilian territory and expand the sugarcane plantation monoculture. Once soils had been completely depleted, new concessions were provided in forest lands, thus creating an expanding and vast network of degraded sugarcane plantation lands. At the end of the 17th Century the Portuguese finally found significant amounts of gold in Brazil and initiated the third economic cycle of the country: gold mining. Agriculture expanded to feed a growing population, and the resulting economic boom destroyed another 3 million ha of forests in the 18th Century [11]. Later, from the mid-19th Century to the beginning of the 20th Century, coffee plantations ended this historical deforestation process in the Atlantic Forest and occupied a major proportion of the southeastern region of the country. To illustrate the severity of deforestation, in the state of São Paulo the remaining Atlantic Forest cover was reduced from 80 to 8% between 1854 and 1973, due largely to coffee expansion to sustain exports to the US and Europe [12]. In sum, all of the historical economic cycles of Brazil occurred at the expense of the Atlantic Forest, the remainder of which is now recognized as one of the top-five global biodiversity Hotspots [13].

After nearly 500 years of massive land use change in the Atlantic Forest, this biome currently has less than 12% of its original forest cover (1.2 million km^2) but houses more than 60% (c.a. 120 million) of the Brazilian population. In addition, the region is responsible for nearly 80% of all Brazilian GDP [14–16]. As a result of an intense process of public land privatization from 1850 forward, with the enactment of the *Lei de Terras* (Law # 601/1850), approximately 90% of the remaining Atlantic Forest is privately held [17]. Thus, the involvement of private landholders in forest restoration initiatives is crucial for both biodiversity conservation and the provisions of ecosystem services in this biome.

3. The Socio-Ecological Context of Habitat Restoration in the Atlantic Forest

Forest restoration initiatives in the Atlantic Forest region started more than 150 years ago. In the late 19th Century, the city of Rio de Janeiro faced water shortages because of the conversion of its original forests/watersheds to agriculture. To reverse this, Emperor Dom Pedro II ordered the planting of thousands of seedlings from 1862 to 1892, and today this forest stands as the Tijuca Forest National Park. Nevertheless, despite this pioneering initiative, it took another one hundred years before forest restoration became truly relevant again in Brazil. Throughout the 20th Century, Brazil enacted a series of legal instruments supporting sustainable use of the forests. These decrees became consistently stronger, eventually obliging farmers to protect key areas for ecosystem services provisioning and requiring private companies to compensate for some of the environmental damage they cause.

The first of these legal instruments was the Forest Code in 1934 (Decree # 23793/1934), which stated that all native forests were of public interest, with an obligation for all rural properties to maintain a certain amount of forest habitat to benefit the entire society. It included a visionary concept of "protective forests", which refer to vegetation that should be conserved to maintain ecosystem services, such as soil retention and water provisioning. However, the law did not establish clearly how much and where, any native forest should be conserved in rural areas. This lack of precision in the law's definitions made enforcement difficult. Thus, in 1965, a revised version of the Forest Code was established (Law #4771/1965), which defined the areas where forests should be preserved—and in some cases restored—to maintain ecosystem services (Areas of Permanent Preservation). It also defined an additional minimum percentage of forest cover for each property

(Legal Reserve), which could be used for sustainable timber harvesting [18]. However, weak environmental governance and the consequent poor compliance with the law hampered the effectiveness of the Forest Code as an instrument to reduce deforestation rates and to foster forest restoration in agricultural landscapes. In the context of large private companies, forest restoration was further stimulated from 1981 forward by the National Environmental Policy (Law # 6938/1981), which established the restoration of degraded lands as part of offsetting policies for companies whose activities cause environmental impacts. This legal instrument boosted forest restoration mainly for mining and hydroelectric companies, which had to compensate for the deforestation caused by their activities.

Following a global trend in reinforcing environmental protection, the Brazilian Federal Constitution established, in 1988, that public authorities should promote restoration of ecological processes in order to guarantee a healthy environment for the Brazilian society. As a consequence, new legal instruments were created to address this concern, resulting in influential support for a restoration initiative in the Atlantic Forest. The Forest Code was further strengthened by a series of complementary laws, which increased the width of Areas of Permanent Protection and the percentage of Legal Reserves in the Amazon. In 1998, the Environmental Crimes Law (Law # 9605/1998) established penal, civil and administrative penalties for individuals and companies responsible for environmental crimes, such as lack of compliance with the Forest Code, and thus designated forest restoration initiatives as a legal obligation for farmers and private companies [19]. From the 2000s onwards, the active role of Public Prosecution relative to environmental laws and the seeking of environmental certification by agricultural companies fostered large-scale restoration programs in many regions of the Atlantic Forest [18]. This trend of continuous strengthening of environmental laws changed in 2012, with the revision of the Forest Code (Law # 12651/2012—now called Law of Native Vegetation Protection) [18]. However, in spite of some environmental setbacks, six million hectares still should be restored or offset by tradable environmental certificates or protected area purchase in the coming years in the Atlantic Forest region in order to comply with the statements of this version of the Code [20].

In addition to the environmental laws mentioned above, which are related to ecological restoration, innovative legal instruments have arisen in recent years to regulate the practice of forest restoration and to increase its socio-ecological benefit, particularly regarding legal compliance, and by providing public funding for restoration [21]. In spite of societal awareness of the need for forest restoration, particularly in the Atlantic Forest, and the large number of legal instruments demanding it, restoration was disorganized, with poor dialogue among the multiple stakeholders and limited incentives for implementation prior to the launch of the AFRP.

4. The Atlantic Forest Restoration Pact: Origin, Motivation and Major Goals

Small-scale forest restoration initiatives have bloomed in the Atlantic Forest region since the 2000s as a result of the growing involvement of (1) environmental NGOs, which moved beyond a perspective of focusing solely on biodiversity conservation to include ecological restoration in their scope of activity; (2) farmers, forced by the Forest Code to restore portions of their lands; and (3) private companies, required to restore native ecosystems by biodiversity offsetting policies and, in some cases, to obtain environmental certification and market benefits [22]. For a variety of reasons, however, the incorporation of these three main groups of stakeholders into restoration activities did not result in a significant expansion of native forests. In the case of NGOs, their main approach to promote forest restoration was to convince farmers to allow restoration of their lands by offering to partially or totally cover the restoration costs in exchange for carbon credits or other benefits. This approach was needed because NGOs usually did not own lands on which to implement restoration projects and most degraded lands in Brazil are found on farmlands. However, most farmers were not interested in forest restoration because they could lose money through the conversion of agricultural land into native ecosystems and/or by investing time and money in the restoration process.

The government also failed to enforce the code related to farmland restoration, further eroding motivation for farmers to participate in restoration efforts, even though such efforts were, theoretically at least, required by law. Consequently, in spite of the existence of the Forest Code, the feeling of impunity caused by very low compliance levels reduced any pressure towards restoration of degraded private lands. This prevented the widespread involvement of farmers in restoration programs, and thus restoration projects led by NGOs or driven by the enforcement of the Forest Code quite often resulted in only very small patches of forests undergoing restoration and with poor integration at the landscape level [23]. Similar challenges arose at the private industry level regarding requirements for compensation and mitigation. Failures and loopholes in public policy, legal enforcement, and monitoring allowed companies to fulfill their legal commitments via the simple establishment of poorly designed tree plantations. Often abandoned after a few years, these plantations did not develop into biologically viable forest stands that could compensate society for the loss of native ecosystems elsewhere [24]. Even environmentally-committed companies interested in implementing effective restoration programs to comply with environmental laws, faced challenges to expanding their programs because of (a) technological constraints, (b) high costs of implementation and maintenance, (c) lack of economic incentives, (d) low ecological effectiveness and (e) weaknesses in the decision-making process [25].

The limitations described above demonstrated that large-scale forest restoration would not be achieved on a case-by-case approach, i.e., by individuals and independent farmers and companies obliged to restore their lands without enough incentives and a pro-active governance approach. This is reinforced by the fact that the decision-making process needed to promote changes in land use and allow restoration is intricately bound up within social, economic, juridical, political, historical and cultural factors [21]. This process cannot therefore be changed on a case-by-case approach or by a group of environmental NGOs. The probability of success is likely to increase however if restoration agents join forces to improve public policies, provide financial incentives for forest restoration while simultaneously discouraging degrading activities, develop appropriate legal instruments to foster and regulate restoration programs, and establish a good governance environment for forest restoration initiatives.

As a result, even though large numbers of conservation NGOs invested a lot of energy and funding to increase the scale of restoration in the biome in order to improve biodiversity conservation and provision of ecosystem services, the results were very disappointing and did not reverse the historical trend of habitat loss and degradation in the biome. Thus the degradation scenario remained the same at a landscape level, with many small and disconnected native forest patches embedded in a matrix dominated by agriculture, with no significant changes that would maintain biodiversity in such human-dominated landscapes [4]. A combination of the need to overcome the constraints preventing the scaling-up of restoration efforts in the Atlantic Forest, and to make certain "structural" transformations to expand forest restoration, stimulated a small group of NGOs and researchers to come together in 2006 to create a diverse coalition to foster large-scale forest restoration in the biome. The group knew that, to be effective, this coalition had to include a confluence of interests and agendas from all key forest restoration actors.

The group developed a plan to move forward and prioritized three steps leading up to the official launching of the AFRP in 2009: (1) engaging and inviting entities from diverse restoration stakeholder groups to join the coalition early in the process, in order to illustrate diversity and improve credibility and impact; (2) developing materials and distributing to members, e.g., a reference book with the lessons learned from 150 years of Atlantic Forest restoration history, restoration methodologies and techniques, a guide for practitioners to implement successful restoration projects, a map of potential restoration areas in the Atlantic Forest, and a website with an online registry system for the main restoration initiatives; and (3) establishing a target for the amount of hectares to be restored. These three steps were critical and challenging, but helped justify the need for such a coalition. Moreover, it was important to demonstrate that a coalition built to achieve large-scale restoration would not conflict with food production, but instead would provide many benefits for farmers and for society in general.

The process of realizing this goal was led by a group of researchers and NGOs. Based on current forest cover and on the target to achieve 30% of forest cover to comply with the Forest Code by 2050, a total of 15 million hectares was established as the restoration target (Figure 1 and Table 1). However, as a result of the new Forest Code in 2012, there was a significant reduction in terms of restoration in the Atlantic Forest biome from about 8.7 to 6.2 million ha [20]. Therefore, to reach the 15 million ha goal the AFRP will need to develop economic restoration models to restore low-productive pasturelands (slope > 15°) that has low opportunity cost (less than US$ 50/ha/year) due to the low productivity and return to the farmers. Another good reason to focus most of the restoration target on low-productivity pasturelands it to avoid competition with food, fuel and fiber production and supply to society.

After initial steps were executed and products generated, the AFRP was officially launched in April 2009, with the goal of restoring 15 million ha of the Brazilian Atlantic Forest by 2050 through the promotion of biodiversity conservation, jobs and income generation, ecosystem services maintenance and provisioning, and by supporting farmers to comply with the Forest Code across the 17 Brazilian states within this biome. To achieve this ambitious goal, the AFRP outlined the following objectives: (a) to establish biologically viable and diverse forests, (b) to enhance the capacity of human-modified landscapes to provide ecosystem services and biodiversity conservation, (c) to develop and implement land use plans that contemplate environmental legislation and minimize negative impacts from economic activities, (d) to build the business case for restoration, and (e) to generate socioeconomic benefits for society.

Figure 1. Fifteen million ha of potential areas for forest restoration mapped in the Atlantic Forest by the Atlantic Forest Restoration Pact in Brazil.

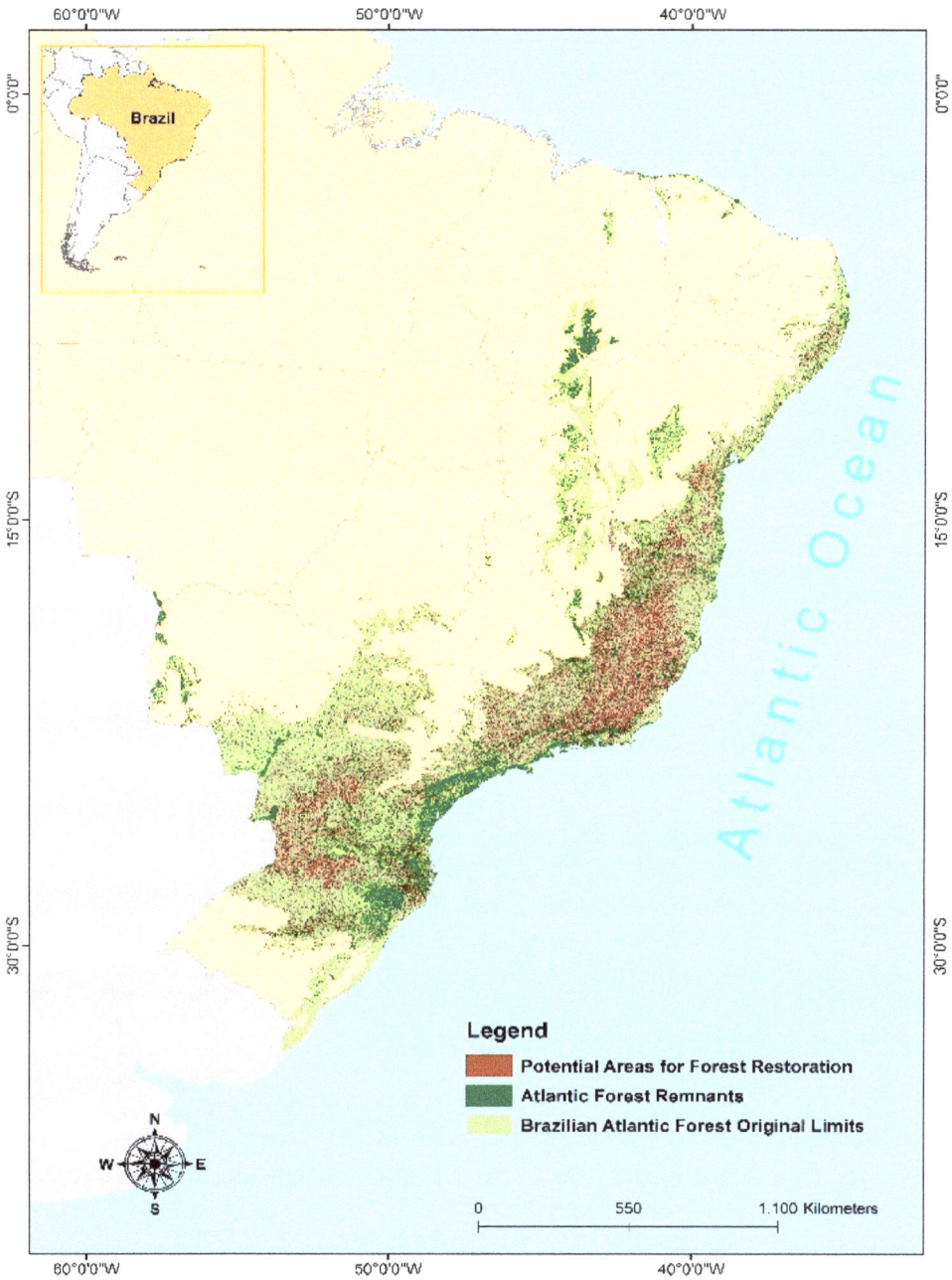

Table 1. Distribution of the original and remaining native vegetation cover in the Atlantic Forest biome, and potential areas for forest restoration mapped by the Atlantic Forest Restoration Pact.

Brazilian region	State	Original area (ha) of Atlantic Forest in each state	Remaining area (ha) of Atlantic Forest in each state (2006)	Potential areas (ha) for restoration in each state (2009)	Number of members *
South	Paraná	19,480,507	4,589,766	2,455,536	22
	Rio Grande do Sul	13,545,367	3,341,227	891	16
	SantaCatarina	9,421,487	3,518,111	1,402,182	8
Central-West	Mato Grosso do Sul	6,287,546	1,123,919	186.453	2
	Goiás	1,050,484	not mapped	not mapped	2
Southeast	Espírito Santo	4,635,982	1,010,845	1,043,374	22
	Minas Gerais	27,660,939	5,646,368	5,648,980	30
	Rio de Janeiro	4,268,142	1,341,634	939.800	70
	São Paulo	16,886,457	3,898,490	2,077,884	131
Northeast	Alagoas	1,508,873	123.879	307	3
	Bahia	18,955,797	3,475,706	2,104,511	33
	Paraíba	639	139	45	6
	Pernambuco	1,804,087	144.411	395	17
	Rio Grande do Norte	314	103	40	2
	Sergipe	1,103,048	145	187.82	6
	Ceará	885.423	not mapped	not mapped	0
	Piauí	2,685,862	not mapped	not mapped	0
Total		131,133,862	28,603,105	17,728,187	370

* Including volunteers, NGOs, private companies, state/local governments and research institutions.

5. Key Governance Structures and Instruments Adopted by the Atlantic Forest Restoration Pact

The AFRP has adopted seven governance structures and instruments, which connect and direct actions and stakeholders towards large-scale forest restoration.

5.1. Members and Coordination Council

The goal of achieving 15 million hectares of restoration and creating biodiversity-friendly and sustainable landscapes, imposes immense legal, technological, ecological, economic, and social challenges and actions. These actions need to be articulated, integrated, coordinated, shared, and aligned between the diverse members of the coalition. The process of becoming a member of the Pact is very simple: An individual or institution representative signs a declaration agreeing with the principles defined in the Protocol of the AFRP. After the institution or individual signs the declaration, the process is assessed by the Executive Secretariat and receives a password to formalize their registration online. Moreover, the process of becoming a member of the AFRP is voluntary and free of charges. The new member is also required to select a level of participation from the following categories: research and dissemination, project executor, public policy formulator, sponsor, seed and seedlings producer, or volunteer. This simple membership process ensures members are aligned with the general objectives and management standards of the AFRP—including the use of restoration technologies and the monitoring protocol—and promotes the exchange of any lessons learned, expertise, and experiences between members. Based on June 2014 figures, the AFRP has 267 members, distributed into four main categories of stakeholders: NGOs, private companies, governments, and research institutions (Figure 2). The majority of members, however, are still NGOs, thus challenging the AFRP to develop a more balanced representation from all categories, in particular the private sector and policy makers.

In order to attract new stakeholders and to mainstream members' involvement in the coalition, a Coordination Council and an Executive Secretariat operate as the central managing body of the AFRP. The Coordination Council is comprised of 21 member institutions (13 NGOs, three research institutions, three governmental agencies and two private companies) and is renewed every two years via an election process decided amongst the members. The roles and responsibilities of the Coordination Council are to establish a strategic plan and a vision for the coalition, and define short and medium-term goals, standards, rules, principles, and policies for the AFRP. The AFRP Coordination Council has a general coordinator and four vice-coordinator chairs representing each stakeholder category.

The role of the Executive Secretariat is to support and oversee the actions of the Council, provide technical and logistical support for capacity building and training courses/workshops for members, and oversee the preparation of primers and technical publications developed by the Pact members. The Secretary is also responsible for updating the website and the database of restoration projects being implemented by the members, engaging new members, and promoting information and experience sharing between members. Aside from the Executive Secretary, all other positions are voluntary and their time is paid by the institutions and organizations they represent as in-kind contributions, making AFRP a low-cost program that promotes the active engagement and participation of its members.

5.2. Regional Units

One of the main challenges faced during the first two years of the AFRP was to unite stakeholders within the 17 states of the Atlantic Forest to get collectively behind the restoration and biodiversity goals. But there is an unbalanced distribution of members within the different geographical regions, which can raise some challenges considering the diversity of ecosystems within the overall 15 million a biome. For example, the Southeast region, in particular the state of São Paulo, has the highest concentration of members, but the majority of potential restoration areas are located in other Brazilian states and regions (Table 1). Thus, it was important to foster and engage the participation of stakeholders from other states and regions to legitimize the AFRP as a national movement. To resolve this, AFRP established "decentralized regional units", which are organized by groups of stakeholders from a given region, and which have the autonomy to establish their own Coordination Board, strategies, and work plans. The first regional unit was created in 2012 in the *Alto São Francisco* region, northeast of Brazil, and coordinated by a member NGO called *Centro de Pesquisas*

154

Ambientais do Nordeste. AFRP is working to stimulate the creation of new regional units in all regions with poor representation and to increase the participation and engagement of a more diverse pool of stakeholders, establishing the AFRP as a truly collective movement within Brazilian society.

Figure 2. Number of members of each affiliation category of the Atlantic Forest Restoration Pact in 2011 and 2014; the percentage values included on the top of the bars represent the increase in the number of members in the period.

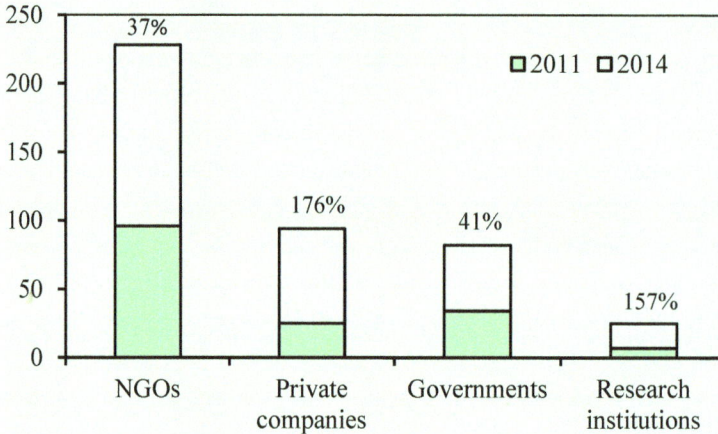

5.3. Working Groups

Because the AFRP could not rely on its own staff to accomplish its goal and objectives, it was necessary to take advantage of the constellation of experts and professionals within the different member institutions. Moreover, the cooperation and participation of different institutions throughout the decision-making process was essential for aligning and integrating members towards a common goal. With this in mind, the AFRP created six working groups (WGs) to coordinate and lead key themes, strategies, and activities. The main functions of the WGs were to (1) provide technical advice to the Coordination Council and the Executive Secretariat in their decision-making process, and (2) find solutions to overcome key barriers for up-scaling restoration initiatives. The six WGs are: Technical-Scientific, Socio-Economic, Fundraising, Public Policies, Information and Knowledge, and Communications and Marketing.

The Technical-Scientific WG is responsible for developing technologies and protocols for ecological restoration, and for building capacity of practitioners and implementers towards large-scale restoration. The Socio-Economic WG aims to transform ecological restoration into an economically viable activity by strengthening different components of the supply chain; evaluating costs, benefits, revenues, and risks associated with forest restoration; and developing innovative financial mechanisms for implementing restoration [15,23]. Two additional goals of this WG are to develop various business cases to attract entrepreneurs interested in investing in restoration, and to conduct research into the social benefits of forest restoration such job creation and income opportunities for farmers and communities. The Fundraising WG is responsible for organizing and approaching potential sponsors, donors, and investors to provide funding support for forest restoration projects, and for the maintenance of the AFRP main structure. The Public Policy WG creates and promotes the adoption of public policies, including legal and economic tools that can contribute to the quality and quantity of forest conservation and restoration in initiatives in the Atlantic Forest. This WG is also responsible for identifying and overturning perverse policies that hamper the advance of restoration in both ecological and socioeconomic terms [21]. In this context, it stimulates the restoration debate within both the AFRP and Brazilian civil society, proposes new regulatory frameworks and public policies, and lobbies for government to approve programs and projects that support ecological restoration. The Information and Knowledge WG is responsible for identifying gaps, organizing information and lessons learned from the projects' performance, mapping priority areas for forest restoration, and ultimately, developing knowledge products and tools that support restoration on the ground. For example, this WG is responsible for geo-referencing relevant information for restoration planning at a landscape level, such as the identification of eligible and suitable areas that supply water to major urban areas and/or for carbon sequestration. Finally, the Communication and Marketing WG is responsible for developing and implementing strategies and actions that disseminate the achievements of the Pact, promote internal and external communication among its members, and increase public awareness of the benefits that come from protecting and restoring the Brazilian Atlantic Forest.

5.4. Training and Capacity Building

Despite over 30 years of scientific background and experience in restoring the Atlantic Forest, the dissemination of this knowledge has occurred only in the last 5–10 years. As a consequence, ecological restoration can be considered a new activity for most of the practitioners, entrepreneurs, policy makers, and other professionals currently engaged in the AFRP. Training and capacity building programs, therefore, are crucial for increasing the scale and quality of restoration projects, and for engaging multiple stakeholders into a common conceptual framework. The AFRP has been building capacity by offering training courses on several topics to empower and increase knowledge among key actors, and to maintain partner engagement and alignment with the goals and objectives of the AFRP.

One of the main achievements of the capacity building program has been the establishment and strengthening of partnerships with both national and international stakeholders. At the national level, many NGOs, private companies, and universities launched training programs in ecological restoration to address the demand for science-based knowledge on all parts of the forest restoration supply chain. Before the AFRP, each stakeholder had to learn by trial-and-error how to establish a nursery facility or to monitor the performance of the project, for example. Currently, several members of the AFRP are promoting training and courses within their area of influence and expertise. At the international level, some of the AFRP members have participated in training and capacity programs in Latin America. They have also disseminated the proven methodologies adopted by the AFRP members at different scales and for different sectors within the restoration supply chain.

5.5. Monitoring Protocol

The credibility of any major forest restoration program is dependent upon the quality of its projects, which can only be demonstrated via a well-designed, cost-effective, and transparent monitoring system with practical indicators. Several large-scale forest "restoration" initiatives have been publicly criticized because they did not meet certain international standards and criteria for ecological restoration. In China, for example, the planting of monoculture plantations with exotic commercial tree species in non-forested habitats was publicized as "forest restoration" [26]. Thus, the challenge of the AFRP is not only to foster the use of methods, techniques, and processes that will increase the likelihood of achieving high-diversity, biologically viable tropical forests, but also to monitor and report if those approaches have succeeded and if the areas being restored achieved the desired biological trajectory. The forest restoration projects included in the AFRP are being monitored through a participatory monitoring protocol that was developed by more than 50 partner institutions over almost three years. The "Monitoring Protocol of the AFRP, which is available online at the AFRP website, includes a set of criteria, indicators, and verifiers for monitoring the ecological, economic, social, and management factors considered critical for the success of any long-term forest restoration project. The main goal of a standard monitoring protocol for assessing the success of restoration projects is to allow comparison among methods, projects and socioeconomic approaches adopted by AFRP members. The results of this monitoring will transform the coalition into a large-scale experiment and provide key findings that will inform the continuous evolution of forest restoration practice and science in the Atlantic Forest.

A new web-based register and monitoring system is being developed to allow members of the AFRP to register their projects and assess their performance against the monitoring protocol indicators, and subsequently make the needed adjustments to ensure the likelihood of project success. Moreover, this system will also allow progress towards the restoration goal of the AFRP and the exchange of any useful lessons learned and other relevant experiences among the members and projects.

6. Main Achievements and Challenges of the Atlantic Forest Restoration Pact

The main challenge of the AFRP as a coalition during the first five years was to engage a critical mass and diversity of stakeholders involved with forest restoration initiatives and to create a favorable governance structure that could achieve the 15-million ha restoration goal by 2050. The engagement and involvement of more than 260 members from different stakeholder groups over the first five years of the AFRP was remarkable, and is rightfully considered a primary achievement. Although the AFRP has been successful in attracting members from various sectors, the representation of each of the four major sectors is not balanced due to an overrepresentation of environmental NGOs. Even though the NGOs make up the majority of the coalition, the other three categories grew significantly between 2011 and 2014, moving AFRP toward a greater balance in the near future. Another limitation of the AFRP has been the uneven geographical distribution of its members, with an overrepresentation of institutions from São Paulo state in the southeast region. Because more than 80% of potential areas for restoration are located outside the state of São Paulo

(Table 1), it is urgent and desirable to have a targeted campaign that attracts members from other states and regions. We expect that the recent and pending creation of decentralized regional units will foster greater participation of underrepresented regions in the coalition.

It has become clear therefore that the success of the AFRP or any other similar coalition depends on the engagement and commitment of its members towards a common vision, goals, and objectives. It also depends on how well-represented the coalition is by all key groups of stakeholders, from the interest, geographic, and representation (e.g., government/business/NGO) perspective. Even though only two private companies are currently represented in the Coordination Council, they represent two of the most demanding sectors for forest restoration: the mining and pulpwood industries. One of these companies is Vale, the biggest mining company in Brazil, an actor highly committed to the vision and goals of AFRP. The second company is Fibria, which is the largest Brazilian pulp producer and which has committed itself to restoring more than 20,000 hectares throughout the next few years. The AFRP recognizes however that the over- or under-representation of sectors and geographical biases reduces the influence and impact of the coalition on national policies and may impose an additional challenge for governing a biome-wide restoration initiative. One solution currently underway is to increase and strengthen regional units, whereby leading members can engage and bring new and more diverse members to the AFRP.

It is also important to strengthen the connection between the AFRP and the agribusiness sector by promoting restoration beyond the conservation agenda, for example, by creating opportunities for investments by landowners and companies. Another important strategy is to create incentives to increase the level of compliance with the new Forest Code and therefore prevent additional changes and/or setbacks to this law.

Almost half of the AFRP's 15-million ha goal was based on the current deficit of Legal Reserves and Areas of Permanent Preservation. Because though the new Forest Code reduces requirements for Forest Restoration, members of AFRP are pursuing strategies and public policies to create new markets and financial incentives to promote "voluntary" restoration projects to meet the 15 million ha goal. These include creating new timber and non-timber forest products markets, promoting payments for ecosystem services (PES), marketing "certified" or environmentally friendly products, and developing more cost-effective approaches to forest restoration. Thus, forest restoration projects must provide a "basket of benefits" for landowners and for the different stakeholders that includes legal, social, environmental, and economic opportunities [23]. The AFRP has begun an initiative to benefit 30 small farmers in the Biodiversity Corridor of Northeastern Brazil, called the "Association of Native Seedlings Producers". The ultimate goal is to convert those farmers into restoration entrepreneurs [23]. The AFRP has also actively participated in the definition of the São Paulo state plan for the implementation of native forests designed for economic exploitation, as part of a bigger plan to create incentives to farmers and increase compliance with the Forest Code.

Policy makers are another stakeholder category that needs better representation in the AFRP, though their engagement has increased as they learned more about how the restoration supply chain can generate green jobs and income to rural communities. A group of AFRP members is actively engaged in discussing these and other public policies with politicians and policy makers. To provide a few examples, AFRP members: (1) laid out the technical and scientific background for protecting and restoring native ecosystems, while also counterbalancing the pressure of the agribusiness lobby when the new Forest Code was being debated; (2) influenced the governor of Pernambuco state to sign an agreement with the coalition to use the AFRP guidelines in forest restoration projects in the state and to offset the degradation caused by the construction of the Suape port, the biggest infrastructure project in Pernambuco state; and (3) achieved a collaboration of the Socio-Economic and Technical-Scientific WGs worked with the National Socio-Economical Development Bank (BNDES) to create financial programs capable of funding restoration projects in the Atlantic Forest.

The AFRP experience in the policy arena has shown that any large-scale forest restoration program cannot rely or depend upon legal compliance as the central motivating factor to achieve restoration targets, given that laws can change depending on the political scenario. Another important lesson is that forest restoration supporters must take an active part in political debates in order to inform and mobilize the public against potential legal setbacks in environmental policy. This requires active lobbying in favor of forest conservation and restoration and good scientific evidence to support these positions. The active involvement of society in political debates concerning forest conservation and restoration is particularly necessary in developing tropical countries, where the pressure to increase food production by replacing natural ecosystems with crops and pasturelands is still very high. In this context, one key part of the discussion about land sparing policies and strategies is that sustainable increases of productivity in cattle ranching could free up land to agriculture. This landscape approach requires coordination, integration, and synergy among agriculture, forest restoration, soil and biodiversity conservation policies, especially to avoid rebound effects where financial gains generated by the productivity improvements could be invested to convert more natural habitats into farm lands.

The production and the widespread use of the AFRP reference book is another major achievement of this coalition. Since most of the technical and scientific information on forest restoration in the Atlantic Forest was spread out among a variety of sources (e.g., scientific articles, books, primers, and proceedings), the organization, synthesis and editing of all relevant information into a reference publication by the Technical-Scientific WG improved the knowledge and awareness of key stakeholders on the science and practice of forest restoration immeasurably. This "state of the art" publication was made available for all members and partners as part of the formal launch of the AFRP and included a set of "technological packages" and "guidelines" on soil preparation, seed and seedling collection and production, planting methods, and management of restoration projects. More than 5000 print copies have been distributed since the launching of the AFRP in 2009, with many more distributed digitally through the AFRP's website [27]. Moreover, several of these protocols and guides have already been adapted and/or are in the process of being adapted to regional ecological and socioeconomic conditions [28]. One of the priorities of the AFRP is to update and translate this reference book into both English and Spanish.

The performance of each WG varies depending on the capacity of the group members. The Technical-Scientific WG, for example, had a rather crucial role at the beginning of the AFRP in creating the main framework and products. This WG also developed tools to ensure the credibility and transparency of restoration efforts being undertaken by the AFRP members, such as guidelines and protocols to help monitor all restoration projects. The Information and Knowledge WG also played a critical role early on by developing the methodology for the potential forest restoration areas map. Some groups, however, have been more limited in their contribution to the AFRP to date. Even though the Fundraising WG have not yet been able to secure a sustainable funding source to maintain the main structure of the coalition and its members, they have made several attempts to mobilize funding and strategic partnerships to support the AFRP. They have been playing a very important role in inserting the AFRP into several global initiatives and have already established some valuable cooperation with the private sector and internationally-financed restoration projects. On the other hand, the Communication and Public Policy WGs are still in search of ways to add value to the coalition and its members. The most recent WG, the Socio-Economic, has been developing and making the business case for forest restoration. Within the next two years, the Socio-Economic WG expects to develop and/or promote innovative economic models to its members that will transform the way restoration is perceived by different sectors and by the public.

The AFRP should strengthen its efforts to make forest restoration a potential economic activity for landowners over the next few years. This medium-term goal is a pre-requisite for scaling-up forest restoration and thus restoring 15-million hectares of forests within the Atlantic forest biome. There is no question that this goal poses some challenges in terms of governance and technology, which will require a significant change in the way the members of the AFRP will be using and deploying their intellectual and human capital in the future.

7. Conclusions

The AFRP is a multi-institutional, multi-partner, bottom-up initiative, which aggregates ideas and actions to achieve large-scale restoration in the Atlantic Forest. By aligning interests and synergies, this cooperation has given a voice to different societal sectors interested in the multiple benefits of forest restoration, allowing the emergence and implementation of a biome-scale restoration initiative. Since its launch in 2009, the AFRP has become a stronger movement despite the economic crises the world has been facing. The governance mechanisms described above are considered fundamental towards achieving this end. Forest restoration for (1) biodiversity persistence, (2) provision of ecosystem services, and (3) socioeconomic development of rural areas would emphasize to multiple sectors of society the wide-ranging benefits of investing in native ecosystems. Not only has the AFRP been expanding its efforts and impacts on the ground, it has also tried to inspire other countries and restoration initiatives to follow a similar approach towards achieving large-scale restoration. The governance mechanisms described above are fundamental for the success of this type of coalition. The AFRP still needs to overcome several challenges, but our experience has shown that a multi-stakeholder network is the clearest way towards realizing large-scale restoration and generating the full range of social, economic, and environmental benefits for the entire society. By sharing the ARFP experience we hope to offer inspiration, lessons and guidance in terms of a general approach, while also acknowledging the multiple challenges that may arise. It is recognized that most tropical biodiversity hotspots lack such a diversity of actors and institutional entities, but the experiences and expertise generated by the ARFP during the last five years can serve as inspiration, providing valuable lessons and models for any large-scale initiative.

Acknowledgments

We wish to thank Manuel Guariguata for the invitation to participate and contribute to this Special Issue. We would also like to thank Simon Dunster and Bethanie Walder for English revision and the members of Atlantic Forest Restoration Pact for the information provided and the inspiration to move forward. This article was published with funding support from Improving the way knowledge on forests is understood and used internationally (KNOW-FOR) program provided by International Union of Conservation of Nature (IUCN). The KNOW-FOR program is funded by the Department for International Development (DFID). The paper expresses views/opinion of authors and not of IUCN/DFID.

Author Contributions

Severino R. Pinto, Marcelo Tabarelli, Felipe Melo, Ricardo G. César and Pedro H.S. Brancalion leaded the preparation and review of the manuscript, while Aurélio Padovesi, Carlos A. Mesquita, Carlos A.M. Scaramuzza, Pedro Castro, Helena Carrascosa, Miguel Calmon and Ricardo Rodrigues actively participated in the development of the Atlantic Forest Restoration Pact and provided general information for the preparation of the manuscript and contributed with the review process.

Conflicts of Interest

The authors declare no conflict of interest.

References and Notes

1. Brancalion, P.H.S.; Melo, F.P.; Tabarelli, M.; Rodrigues, R.R. Biodiversity persistence in highly human modified tropical landscapes depends on ecological restoration. *Trop. Conserv. Sci.* **2013**, *6*, 705–710.
2. Mcconnachie, M.M.; Cowling, R.M.; Shackleton, C.M.; Knight, A.T. The challenges of alleviating poverty through ecological restoration: insights from South Africa's "Working for Water" Program. *Restor. Ecol.* **2013**, *21*, 544–550.
3. Aronson, J.; Alexander, S. Ecosystem restoration is now a global priority: Time to roll up our sleeves. *Restor. Ecol.* **2013**, *21*, 293–296.
4. Melo, F.P.L.; Arroyo-Rodríguez, V.; Fahrig, L.; Martínez-Ramos, M.; Tabarelli, M. On the hope for biodiversity-friendly tropical landscapes. *Trends. Ecol. Evo.* **2013**, *28*, 462–468.
5. Menz, M.H.M.; Dixon, K.W.; Hobbs, R.J. Hurdles and opportunities for landscape-scale restoration. *Science* **2013**, *339*, 526–527.
6. Mcqueen, C.; Noemdoe, S.; Jezile, N. The working for water programme. *Land Use Water Resour. Res.* **2001**, *1*, 1–4.
7. Arriagada, R.A.; Ferraro, P.J.; Sills, E.O.; Pattanayak, S.K.; Cordero-Sancho, S. Do payments for environmental services affect forest cover? A farm-leve evaluation from Costa Rica. *Land Econ.* **2012**, *88*, 382–399.

8. Mittermeier, R.; Turner, W.; Larsen, F.; Broks, T.; Gascon, C. *Global Biodiversity Conservation: The Critical Role of Hotspots*; Springer Berlin Heidelberg: Berlin, German, 2011; pp. 3–22.
9. Brancalion, P.H.S.; Viani, R.A.G.; Calmon, M.; Carrascosa, H.; Rodrigues, R.R. How to organize a large-scale ecological restoration program? The framework developed by the Atlantic Forest restoration pact in Brazil. *J. Sustain. For.* **2013**, *32*, 728–744.
10. Rodrigues, R.R.; Lima, R.A.F.; Gandolfi, S.; Nave, A.G. On the restoration of high diversity forests: 30 years of experience in the Brazilian Atlantic Forest. *Biol. Conserv.* **2009**, *142*, 1242–1251.
11. Dean, W. With Broadax and Firebrand: The Destruction of the Brazilian Atlantic Forest; University of California Press: London, UK, 1995; p. 485.
12. Victor, M.A.M.; Cavalli, A.C.; Guillaumon, J.R.; Filho, R.S. *Cem Anos de Devastação*; Brazilian Environmental Ministry: Brasília, Brazil, 2005; p. 72.
13. Laurance, W.F. Conserving the hottest of the hotspots. *Biol. Conserv.* **2009**, *142*, 1137.
14. Tabarelli, M.; Roda, S.A. Oportunidade para o Centro de Endemismo Pernambuco. *Natur. Cons.* **2005**, *3*, 22–28.
15. Brancalion, P.H.S.; Cardozo, I.V.; Camatta, A.; Aronson, J.; Rodrigues, R.R. Cultural ecosystem services and popular perceptions of the benefits of an ecological restoration project in the Brazilian Atlantic forest. *Restor. Ecol.* **2014**, *22*, 65–71.
16. Calmon, M.; Brancalion, P.H.S.; Paese, A.; Aronson, J.; Castro, P.; Silva, S.C.; Rodrigues, R.R. Emerging threats and opportunities for large-scale ecological restoration in the Atlantic Forest of Brazil. *Restor. Ecol.* **2011**, *19*, 154–158.
17. Sparovek, G.; Berndes, G.; Klug, I.L.F.; Barreto, A.G.O.P. Brazilian agriculture and environmental legislation: status and future challenges. *Environ. Sci. Technol.* **2010**, *44*, 6046–6053.
18. Garcia, L.C.; Santos, J.S.; Matsumoto, M.; Silva, T.S.F.; Padovezi, A.; Sparovek, G.; Hobbs, R.J. Restoration challenges and opportunities for increasing landscape connectivity under the New Brazilian Forest. *Act. Natur. Cons.* **2013**, *11*, 181–185.
19. Durigan, G.; Melo, A.C.G. Panorama das políticas públicas e pesquisas em restauração ecológica no estado de São Paulo. In *Conservación de la Biodiversidad en las amÉricas: Lecciones y Recomendaciones de Política*; Figueroa, E., Ed.; Universidad de Chile: Santiago, Chile, 2011; pp. 355–387.
20. Soares-Filho, B.; Rajão, R.; Macedo, M.; Carneiro, A.; Costa, W.; Coe, M.; Rodrigues, H.; Alencar, A. Cracking Brazil's Forest Code. *Science* **2014**, *344*, 363–364.
21. Aronson, J.; Brancalion, P.H.S.; Durigan, G.; Rodrigues, R.R.; Engel, V.L.; Tabarelli, M.; Torezan, J.M.D.; Gandolfi, S.; Melo, A.C.G.; Kageyama, P.Y.; *et al.* What role should government regulation play in ecological restoration? Ongoing debate in São Paulo State, Brazil. *Restor. Ecol.* **2011**, *19*, 690–695.
22. Wuethrich, B. Reconstructing Brazil's Atlantic Rainfores. *Science* **2007**, *315*, 1070–1072.
23. Brancalion, P.H.S.; Viani, R.a.G.; Strassburg, B.B.N.; Rodrigues, R.R. Finding the money for tropical forest restoration. *Unasylva* **2012**, *63*, 41–50.

24. Maron, M.; Hobbs, R.J.; Moilanen, A.; Matthews, J.W.; Christie, K.; Gardner, T.A.; Keith, D.A.; Lindenmayer, D.B.; Mcalpine, C.A. Faustian bargains? Restoration realisities in the context of biodiversity offset policies. *Biol. Conserv.* **2012**, *155*, 141–148.

25. Tollefson, J. Brazil revisits forest code. *Nature* **2011**, *476*, 259–260.

26. Xu, J. China's new forests aren't as green as they seen. *Nature* **2011**, *447*, 371.

27. Rodrigues, R.R.; Brancalion, P.H.S.; Isernhagen, I. *Pacto Pela Restauração da Mata Atlântica: Referencial dos Conceitos e Ações de Restauração Florestal*; Instituto BioAtlântica: São Paulo, Brazil, 2009.

28. Rodrigues, R.R.; Gandolfi, S.; Nave, A.G.; Aronson, J.; Barreto, T.E.; Vidal, C.Y.; Brancalion, P.H. Large-scale ecological restoration of high-diversity tropical forests in SE Brazil. *For. Ecol. Manag.* **2011**, *261*, 1605–1613.

China's Conversion of Cropland to Forest Program for Household Delivery of Ecosystem Services: How Important is a Local Implementation Regime to Survival Rate Outcomes?

Michael T. Bennett, Chen Xie, Nicholas J. Hogarth, Daoli Peng and Louis Putzel

Abstract: China's Conversion of Cropland to Forests Program (CCFP) is the world's largest afforestation-based Payments for Ecosystem Services (PES) program, having retired and afforested over 24 million ha involving 32 million rural households. Prior research has primarily focused on the CCFP's rural welfare impacts, with few studies on program-induced environmental improvements, particularly at the household level. In this study, data from a 2010 survey covering 2808 rural households from across China was analyzed using an interval regression model to explain household-reported survival rates of trees planted on program-enrolled cropland. In addition to household-level factors, we explore the influence of local conditions and institutional configurations by exploiting the wide diversity of contexts covered by the data set. We find that households with more available labor and more forestry experience manage trees better, but that higher opportunity costs for both land and labor have the opposite effect. We also find that the local implementation regime- e.g., the degree of prior consultation with participants and regular monitoring - has a strong positive effect on reported survivorship. We suggest that the level of subsidy support to participating households will be key to survivorship of trees in planted CCFP forests for some time to come.

Reprinted from *Forests*. Cite as: Bennett, M.T.; Xie, C.; Hogarth, N.J.; Peng, D.; Putzel, L. China's Conversion of Cropland to Forest Program for Household Delivery of Ecosystem Services: How Important is a Local Implementation Regime to Survival Rate Outcomes? *Forests* **2014**, *5*, 2345-2376.

1. Introduction

Catastrophic drought and flooding during 1997–1998 in China's two major river basins–the Yellow and Yangtze Rivers—catalyzed an important turning point in China's forest policy. The Yellow River witnessed a historic dry-out in 1997, whereby it did not reach the sea for an unprecedented 267 days. This was followed in the summer of 1998 by major floods in the Yangtze River Basin and the Songhua and Nen rivers in Northeast China, which are estimated to have claimed 3,000–4,000 lives and caused more than US$12 billion in damages and lost production, including the loss of some five million hectares of crops [1–6]. In response, the central government launched a portfolio of programs - referred to as the "Six Key National Forestry Programs" - aimed at shifting the focus of the forest sector from its primary emphasis on extractive timber production to a more balanced approach involving concerted efforts to rehabilitate/restore and more sustainably manage forests for the ecological services they provide [3,7]. The largest, most important and most innovative of these programs is arguably the "Conversion of Cropland to Forests Program" (CCFP) [1].

Also known as "Grain-for-Green" or the "Sloping Land Conversion Program," the CCFP is the world's largest afforestation-based Payments for Ecosystem Services (PES) program. It involves over 32 million rural households and an investment of more than US$42 billion (up until 2013), and has over 27 million ha of land enrolled [8]. Given the program's extent and size, and thus its potentially important impacts on the rural economy, previous work on the CCFP has primarily explored its rural welfare impacts—in terms of household income effects, labor allocation and structure of production [9–14]—as well as general implementation of the program, including land targeting and program cost effectiveness [1,10,15–17]. Some work has also looked at the program's potential impacts on grain output from reductions in crop area, which has been a recent concern of policymakers [18–20].

While such work has been important for gauging the long-term economic sustainability of CCFP, and by implication its environmental outcomes, it has had relatively little to directly say about program-induced provision of ecosystem services, in particular at the household level. Understanding how program incentives and other economic and social factors influence and induce rural households to provide targeted forest ecosystem services is critical for ensuring the CCFP's success, and for gauging the program's future. However, with the exception of Bennett and colleagues [8], work to date examining the CCFP's environmental dimensions has either been primarily qualitative in nature, has estimated future environmental benefits based on landscape scale models under hypothetical scenarios, or has used stated choice methods to estimate public willingness to pay for the program [11,16,21].

To get at the question of program-induced household delivery of targeted forest ecosystem services, we use a large 2010 rural household data set collected from across China to examine the factors associated with the survival rate of household/program-planted trees. While clearly not quantifying the multiple ecosystem services provided by trees and forests, given the current availability of data at the national level the survival rate of program-planted trees is nonetheless an appropriate measure for examining the impacts of program incentives on household ecosystem service provision; it is the key indicator by which household CCFP implementation is evaluated and subsidy delivery determined. In other contexts, the survival rate of trees (in particular during the establishment phase, which is a 3–5 year period from when seed or seedlings are planted) is also used in program assessments [22,23].

The richness of the survey data set allows us to examine the impact of household characteristics, local institutions and socioeconomic context on survival rates. Implementation of the CCFP often varies significantly by locale, as is generally the case with many central government policies in China [24,25], and we exploit this to examine how the different configurations in a local *de facto* implementation regime are associated with reported survival rates.

2. The Conversion of Cropland to Forests Program

The CCFP was launched in 1999 via piloting in the three western provinces of Sichuan, Gansu, and Shaanxi, followed by full-scale implementation that expanded the program to 25 provinces by 2002 [26]. The policy originally targeted reducing soil erosion and flooding, but this has since been expanded to also emphasize local economic development and poverty alleviation, in line with an evolving national policy discourse [5,26]. To date, the program has converted 27.55 million ha of land into primarily tree-based plantations, of which 9.06 million ha (an area the size of Portugal) is retired and afforested cropland, 15.80 million ha is formally "wasteland" (also translated as "barren land", an official land category in China that includes marginal or sloping land that is deemed suitable to be developed into cropland), and 2.68 million ha is remote mountainous areas that have been sealed off to allow natural regeneration [8].

Significantly, the program directly pays participating rural households to retire and afforest or plant vegetative cover on their sloping or marginal cropland, and to ensure that these planted trees/grasses survive. Subsidy levels and duration depend on region and whether grasses, "ecological trees" (which can be either timber crops for harvesting or trees for providing ecological services) or "economic trees" (orchard crops, trees with medicinal value or trees that produce other types of NTFPs) are planted. According to the original 2003 plan, subsidy payments included a one-time fee of 750 CNY/ha for saplings or seeds, an annual living allowance of 300 CNY/ha, and an annual grain/cash subsidy that was differentiated according to whether the participating household is in Yangtze River or Yellow River watershed regions (1 USD = 8.28 CNY based on 2003 yearly average exchange rates). The subsidy rates were, and still are, the same irrespective of the type of tree or grass planted, with only the length of payments differing (see Table 1) [27]. The original plan also generally required that participating households afforest a roughly equal area of "wasteland," though the degree to which this was adhered to has varied by locale [1]. Phase II of the program, which commenced in 2007, has doubled the subsidy period, continuing both the living allowance and subsidies, with the latter now wholly in cash [27] (Table 1).

The CCFP targets a range of bundled forest ecosystem services, including timber, carbon sequestration, biodiversity habitat, landscape amenities and watershed services [1]. While program indicators for delivery of these have included land area afforested, types of land afforested (e.g., sloped and marginal land) and choice of trees to plant, the survival rates of program-planted and managed trees has been the most explicit program indicator utilized for evaluating service provision. During the pilot phase (1999–2002), subsidy delivery to households was stipulated to be conditional on achieving a survival rate for trees planted on enrolled land of 85% for Yangtze River basin regions, and 70% for Yellow River basin regions. This has since been revised to a nationwide standard of 75% for full-scale implementation [1]. However, this rate has varied by locale, often due to the tension faced by local governments between maintaining enthusiasm for the program and ensuring its rural welfare goals (*i.e.*, by making sure households get program subsidies), while also achieving program environmental goals (*i.e.*, by incentivizing achievement of survival rate targets by withholding a share of subsidies for sub-par outcomes) [24].

Table 1. The Conversion of Cropland to Forests Program subsidy rates.

Subsidy Component	Subsidy Rates			Duration		
	Yangtze River Watershed and South China	or	Yellow River Watershed and North China	Ecological Forests	Economic Forests	Grasses
PILOT PHASE (1999–2001): 412 counties in 20 provinces [1]						
	ONE-TIME PAYMENT					
Sapling/seedling Fee	750 CNY/ha			One-time, upon enrollment		
	ANNUAL PAYMENTS [2]					
(i) Living Allowance (Cash)	300 CNY/ha			Payment length as yet undermined.		
(ii) Annual Subsidy (Grain)	2,250 kg/ha (*i.e.*, 3,150CNY/ha)	or	1,500 kg/ha (*i.e.*, 2,100 CNY/ha)			
PHASE I-Full-scale implementation (2002–2007): 2500 counties in 25 provinces [3]						
	ONE-TIME PAYMENT					
Sapling/seedling Fee	750 CNY/ha			One-time, upon enrollment		
	ANNUAL PAYMENTS					
(i) Living Allowance (Cash)	300 CNY/ha			8 years	5 years	2 years
(ii) Annual Subsidy (Grain or cash equivalent value)	2,250 kg/ha (*i.e.*, 3,150CNY/ha)	or	1,500 kg/ha (*i.e.*, 2,100 CNY/ha)			
PHASE II (2008–2016)						
	CONTINUED ANNUAL PAYMENTS					
(i) Living Allowance (Cash)	300 CNY/ha			+8 years	+5 years	+2 years
(ii) Annual Subsidy (Cash only)	1,575 CNY/ha	or	1,050 CNY/ha			

Source: State Forestry Administration, 2003; State Council, 2007; Average pilot phase exchange rate was 1 USD = 8.28 CNY; average phase I exchange rate was 1 USD = 8.10 CNY; average phase II exchange rate was 1 USD = 6.52 CNY; [1] the pilot phase lasted three years; it was launched in 1999 in Sichuan, Gansu and Shanxi province. In 2000, the pilot extended to 188 counties of 17 provinces, and in 2001, it further expanded to some 400 counties of 20 provinces; [2] Subsidy durations were not decided during the pilot phase; upon full-scale implementation, the length of time subsidies had already been delivered during the pilot phase was counted towards the formal subsidy lengths stipulated; [3] Though a formal shift to cash-only subsidies was not stipulated until State Council (2007), State Council (2004) provided standards for converting grain to cash subsidies at the rate of 1.4 CNY/kg, since many locales paid subsidies fully in cash by that time.

3. Empirical Strategy

Using a 2010 rural survey, we utilize household-reported survivorship of program-planted trees on the household CCFP-enrolled land to explore how household characteristics and local implementation regime are associated with household-level delivery of CCFP-targeted ecosystem services. In particular, we analyze household responses to the following question asked in the survey:

What has been the survival rate of the trees on your CCFP enrolled land thus far?

   A. Over 90%;

   B. 70%–90%;

   C. 40%–70%;

   D. 10%–40%;

   E. Less than 10%.

In the sample selected for analysis, 19% of household reported ranges of 70% and under (answers C, D or E in the question above), 40% of households reported the 70%–90% range, while the remaining 41% of households reported over 90% (see Figure 1 below).

Figure 1. Regional distribution of survival rates of household conversion of cropland to forests program planted trees.

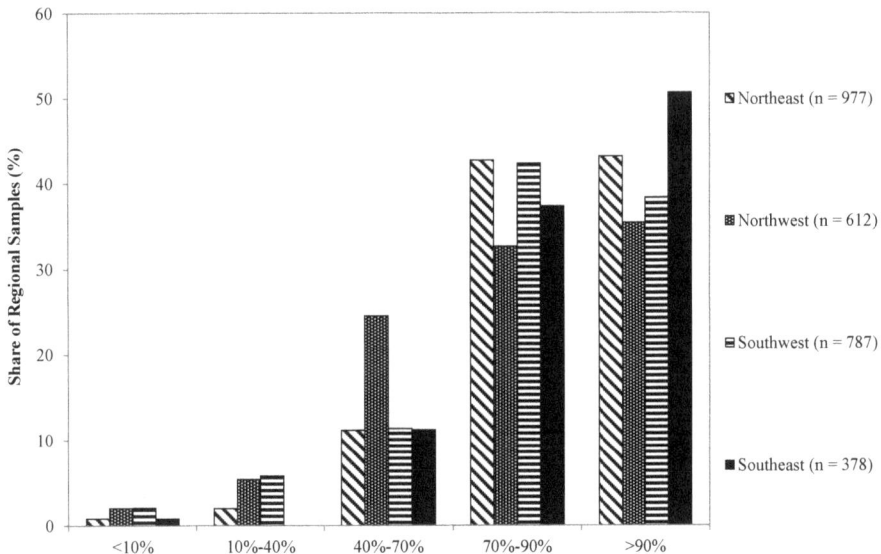

Note: The regional classifications in Figure 1 are not the SFA's formal delineations, but rather they are the regional indicator variables constructed for this analysis; "Northeast" consists of Beijing, Hebei, Shanxi, Inner Mongolia, Liaoning, Jilin and Heilongjiang provinces; "Northwest" consists of Tibet, Shaanxi, Gansu, Qinghai, Ningxia and Xinjiang provinces; "Southwest" covers Guangxi, Hainan, Chongqing, Sichuan, Guizhou and Yunnan provinces; and "Southeast" consists of Anhui, Jiangxi, Hubei and Hunan provinces.

The data used for this analysis comes from a unique, one-time cross-sectional rural survey collected in 2010 and spanning much of China. The survey encompasses 2808 rural household CCFP participants, from 419 villages across 228 townships, 132 counties and 24 provinces, and was collected by 125 Beijing Forestry University (BFU) students upon returning to their home-towns/villages during the annual spring festival (see Figure 2 below). Student enumerators volunteered to participate in the survey, and numbered 159 at the beginning of the project and training. At the end of the fieldwork, 125 enumerators were able to provide completed surveys, from which the data for this analysis has been complied. Overall, the survey was well implemented, with strong support of the BFU student union and in close collaboration with the State Forestry Administration's National Forest Economics and Development Research Center (FEDRC), the agency responsible for monitoring and assessing the socioeconomic and environmental impacts of the CCFP and other key forestry programs. Arguably the most representative data set yet available for evaluating the CCFP, it is the result of one of a number of innovative initiatives and approaches utilized by the FEDRC to monitor this very large and regionally diverse program.

Figure 2. Conversion of cropland to forests program provincial coverage and survey data sample counties.

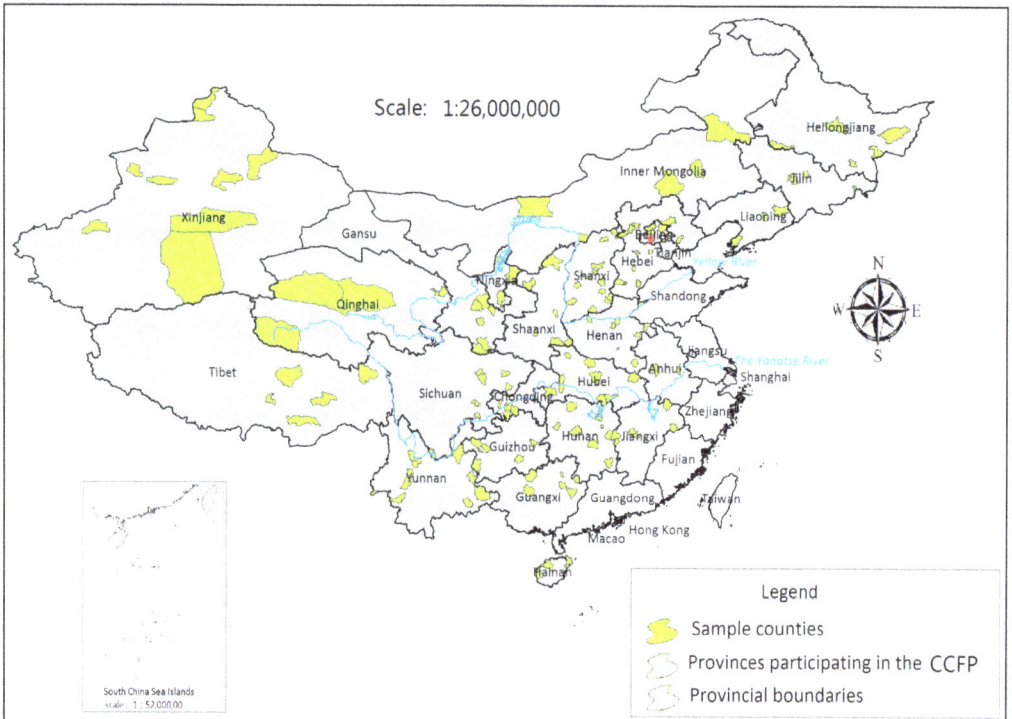

Map produced by Wang Jiang (Beijing Forestry University and FEDRC).

Students received basic enumerator training about data collection and household sampling methods prior to collecting survey data in their home regions in villages participating in the CCFP. Households were asked a range of questions regarding socioeconomic characteristics, income, employment, dependence on agriculture, forestry experience, local CCFP implementation regime, the characteristics of other related government policies in the locale, and a range of variables capturing program outcomes and household views regarding program strengths, weaknesses and induced local changes. Sample households were mostly randomly selected using official government lists of household names (Random sampling was likely not applied in all cases by students, though it is not possible to identify where this occurs in the data set). The completed, hand-written surveys were sent to BFU where data was entered into Excel before being handed over to the FEDRC for analyses. Of the 2808 surveys collected, 2635 households were selected for inclusion into the sample for analysis. Households excluded included: 114 with numerous missing explanatory variables, 30 with extremely large enrolled land area relative to the sample average, 122 that planted grasses on enrolled land, and 21 with missing responses to the survival rate question.

3.1. Model

Given the nature of the dependent variable, an interval regression model is used for the analysis. An interval regression is a generalized form of limited dependent variable models wherein the underlying dependent variable is unobserved, but the interval within which it falls is. Let the actual (*i.e.*, unobserved) survival rate be characterized as $y^* = X\beta + \varepsilon$, where X is a vector of explanatory variables, β the vector of parameters for these, and $\varepsilon \sim N(0, \sigma^2 I)$. Let y_i be the observed survival rate ranges within which y^* falls in the data set, whereby:

$$y_i = \begin{cases} y_1 & if \ \theta_{1LB} \le y^* \le \theta_{1UB} \\ y_2 & if \ \theta_{2LB} < y^* \le \theta_{2UB} \\ \quad \vdots \\ y_m & if \ \theta_{mLB} < y^* \le \theta_{mUB} \end{cases} \tag{1}$$

Since $y^* \in (0,100)$, $\theta_{1LB} = 0$ and $\theta_{mUB} = 100$. The log likelihood function for the model is

$$\ln L = \sum_{n=1}^{N} \sum_{i=1}^{m} \pi_{ni} \cdot \log \left\{ \Phi \left(\frac{\theta_{iUB} - x_n\beta}{\sigma} \right) - \Phi \left(\frac{\theta_{iLB} - x_n\beta}{\sigma} \right) \right\} \tag{2}$$

where $\Phi(\cdot)$ is the standard normal cdf, and:

$$\pi_{ni} = \begin{cases} 1 \ if \ y_i \ is \ the \ interval \ observed \ for \ household \ n, \\ \quad\quad\quad 0 \ otherwise. \end{cases} \tag{3}$$

Of interest is the degree to which household characteristics, local implementation regime, and other related conditions are associated with the survival rates of household CCFP-planted trees. To explore this, the unobserved survival rate is modeled as:

$$y^* = \alpha + \gamma\beta^Y + X^H\beta^H + X^I\beta^I + X^C\beta^C + \varepsilon \tag{4}$$

Where α is the intercept term, γ is a vector of regional indicator variables, X^H is the vector of household characteristics, X^I is a vector of local CCFP implementation regime characteristics, X^C is a vector additional relevant local characteristics, $\{\beta^{\gamma}, \beta^{H}, \beta^{I}, \beta^{C}\}$ are the associated parameters and ε the error term.

3.2. Explanatory Variables

A range of model specifications were explored to examine how local institutional arrangements governing or related to CCFP implementation are associated with survival rates. First, a "naïve" model was estimated wherein local institutional regime and context is controlled for via household survey responses regarding this. Since such variables are likely correlated with both observed and unobserved household characteristics (e.g., aspects of human and social capital such as level of education and access to information, and psychological factors associated with household views on program implementation and outcomes) and therefore biased estimators of the actual local implementation regime, additional characterizations of the model were explored in which the local institutional variables were constructed from averages across households in the same locale.

Table 2 details sample household characteristics used as explanatory variables in the model. These falls into three categories: "Household Socioeconomic Characteristics"; "CCFP implementation"; and "Other Policy Impacts." Household socioeconomic characteristics were selected to capture the effects of household welfare and social capital (such as whether the household considers itself to be poor or rich within the village, and it is a member of an ethnic minority group), level and structure of income (whether agriculture is the main source of income, and the share of household labor that is "migrant," which usually means long-term), and rough measures of the household's human capital and experience (such as pre-CCFP per capita cropland and managed forest area, whether the household has worked before in afforestation, and whether the household has highly sloping cropland).

Table 2. Explanatory variables–household characteristics, 2010.

Sample Household Characteristics	Full Sample ($n = 2656$)			
	Mean	SD	Min	Max
Household Socioeconomic Characteristics (X^{H})				
HH Labor Population (>15 Years Old)	3.07	1.39	0	16
Respondent Has High-School or Above Education (1 = Yes)	0.17	0.38	0	1
Elementary School or Middle-School Education (1 = Yes)	0.73	0.45	0	1
Has No Education or "Other" (1 = Yes)	0.09	0.28	0	1
Share of HH Labor that is "Migrant"	0.35	0.35	0	1
HH PC Crop Area Pre-CCFP (ha)	0.2	0.24	0	2.1
HH PC Forest Area Pre-CCFP (ha)	0.06	0.37	0	13.3
Agriculture is HH's Main Income Source (1 = Yes)	0.60	0.49	0	1
HH is Poor in the Village (1 = Yes)	0.15	0.36	0	1
HH is Rich in the Village (1 = Yes)	0.10	0.30	0	1
HH Members Worked Before in Afforestation (1 = Yes)	0.11	0.32	0	1
HH Has cropland > 25 Degrees (1 = Yes)	0.42	0.49	0	1
Ethnic Minority? (1 = Yes)	0.17	0.37	0	1
CCFP Implementation (X^{H}, X^{I} in the "naïve" model)				
Years in CCFP	6.21	1.96	2	10
HH CCFP Land Has Been Inspected Before (1 = Yes)	0.78	0.42	0	1
HH CCFP Land Is Formally Registered Agricultural Land (1 = Yes)	0.77	0.42	0	1

Table 2. *Cont.*

Sample Household Characteristics	Full Sample (*n* = 2656)			
	Mean	SD	Min	Max
CCFP Implementation (X^H, X^I in the "naïve" model)				
HH Has Other CCFP Afforestation Responsibilities (1 = Yes) *	0.25	0.43	0	1
HH "Probably/Partially Understands" the Policy	0.50	0.50	0	1
HH "Doesn't Understand" the Policy	0.17	0.38	0	1
HH "Knows" Their Responsibilities under CCFP **	0.85	0.36	0	1
HH Receives Subsidies in Cash	0.22	0.41	0	1
HH Receives Subsidies via Smart Card	0.69	0.46	0	1
CCFP Management Type is "Large HH" or "Large HH + Company"	0.08	0.27	0	1
Timber Trees (1 = Yes)	0.41	0.49	0	1
Orchard Trees (1 = Yes)	0.39	0.49	0	1
Joint Type of Tree (e.g., Bamboo) (1 = Yes)	0.10	0.30	0	1
Shrubs (1 = Yes)	0.09	0.29	0	1
Had to Change Trees Types on CCFP Land? (1 = Yes)	0.13	0.33	0	1
CCFP Land Main Type Is Sloped (1 = Yes)	0.59	0.49	0	1
CCFP Land Main Type Is Desertified (1 = Yes)	0.14	0.35	0	1
CCFP Plots Average Distance from Home (km)	1.86	3.42	0	35
CCFP Land Was Hit by Disaster (1 = Yes)	0.08	0.28	0	1
Intercropping on CCFP Land (1 = Yes)	0.19	0.39	0	1
HH Per Capita CCFP Land Area (ha)	0.12	0.2	0	2
Average Yield of CCFP Land was High, Pre-CCFP (1 = Yes)	0.08	0.28	0	1
Other Policy Impacts (X^H, X^I in the "naïve" model)				
HH Forest Land is Collectively Managed	0.07	0.26	0	1
Received Forest Certification (part of collective forest sect reforms)	0.58	0.49	0	1

* This can be either wasteland afforestation, or "closed-mountain" afforestation; ** these are indicated as "Forest Management" and "Afforestation" responsibilities.

CCFP implementation regime variables were selected to capture the effects of local program implementation regime (e.g., inspections, subsidies, enrollment intensity, and program land management aspects), ecological dimensions affecting survival rate outcomes (e.g., broad categories of tree types planted, enrolled land characteristics and pre-CCFP yield, whether replanting was necessary), and household level of understanding of program goals and responsibilities. Finally, under "Other Policy Impacts", two variables were also included to capture the impacts of the degree to which in the locale in question collective forest sector reforms were being implemented—which are ongoing reforms in China's southern collective forest sector whereby forest management rights are gradually being devolved to local communities and households.

To improve identification of local institutional characteristics, within-village averages across household responses for villages with at least 15 survey households were constructed as explanatory variables. While a relatively unified CCFP implementation regime possibly extends up to the county level in many locales across China, this is too large a unit of analysis for statistical identification of institutional effects and regional characteristics given the small available household samples per county in this data set, as well as the fact that such samples are generally clustered in particular villages rather than evenly spread throughout the county. Furthermore, the village is generally the most appropriate unit of analysis for institutional impacts on household behavior in rural China. For example, though national law stipulates set rights and tenure configurations for agricultural land, Rozelle and colleagues [28] find that in reality significant heterogeneity in *de facto* agricultural land rights exists across China at the village level, with this being the result of village governance factors.

Although not an ideal sample size, 15 was chosen as a cut-off point based on a balance between statistical rigor and minimal reduction in the total sample size. The household-level sample sizes for the samples constructed thus are 1675 and 1785, respectively. Restricting the sample to village or township clusters of higher than 15 households quickly reduced the overall sample sizes for analysis. To examine the degree to which such variables produced robust parameter estimates, specifications using similarly constructed township-level variables were also examined for comparison.

Table 3 details the methodology used for construction of the village-level variables from household survey responses, as well as the descriptive statistics of the variables so constructed. In total, 92 villages had at least 15 survey households from which to construct variables, while 93 townships had such. Depending on the particular institutional or economic characteristic being captured, the mean, median or maximum of household-level responses for the village sample were used as appropriate. Sample means were used to represent a village's propensity to have a particular characteristic or condition while sample medians of household-level 0/1 indicator variables were used as discrete indicators of a village condition or institution as appropriate. The number of years the village had been implementing the CCFP is estimated as the maximum number from the household sample in that village.

Finally, to control for systematic differences in regional ecological and economic conditions that could affect survival rates, regional indicator variables were constructed to divide the sample roughly into the four regions: "Northeast" (Beijing, Hebei, Shanxi, Inner Mongolia, Liaoning, Jilin and Heilongjiang provinces), "Northwest" (Tibet, Shaanxi, Gansu, Qinghai, Ningxia and Xinjiang provinces), "Southwest" (Guangxi, Hainan, Chongqing, Sichuan, Guizhou and Yunnan provinces) and "Southeast" (Anhui, Jiangxi, Hubei and Hunan provinces). For the sample analyzed, 36% is in the northeast, 21% is in the northwest, 29% is in the southwest, and 14% is in the southeast.

4. Results and Discussion

Table 4 below presents the interval regression model results. Specification (1) only includes household characteristics, while (2), (3) and (6) are the "naïve" models wherein only household-level responses are used as indicators of local implementation regime. Specifications (4)–(5) and (7)–(8) utilize the constructed village and township-level variables, respectively, to capture local institutional and other characteristics. Specifications (3) and (6) were included to examine in what way restricting estimation to the village and township samples affects the parameter estimates of the "naïve" model.

Table 3. Explanatory variables—constructed village institutional, conversion of cropland to forests program implementation and other characteristics, 2010.

Constructed Village Characteristics	Statistic Used [+]	Interpretation	Village Sample (n = 92)				Township Sample (n = 93)			
			Mean	SD	Min	Max	Mean	SD	Min	Max
CCFP Implementation Regime (X^I)										
Number of Years Village has Been Implementing CCFP	Max	(A) Village began implementing CCFP since at least the earliest enrollment year reported by village households in the sample.	7.24	1.79	2	10	7.38	1.79	2	10
... = Share of households in the village										
... with CCFP land inspected	Mean	(B) Capturing village propensity to have this characteristic, measured as share of village households in the sample reporting this.	0.81	0.29	0	1	0.81	0.29	0	1
... with intercropping on CCFP land			0.16	0.24	0	1	0.17	0.24	0	1
... that "understand" the policy			0.35	0.32	0	1	0.36	0.33	0	1
... that "probably/partially understand" the policy			0.51	0.31	0	1	0.49	0.31	0	1
... that "do not understand" the policy			0.15	0.24	0	0.93	0.15	0.25	0	0.93
... "know their responsibilities" under CCFP [++]			0.86	0.22	0.03	1	0.87	0.20	0.03	1
... with other CCFP afforestation responsibilities [+++]			0.27	0.37	0	1	0.30	0.38	0	1
... that received full subsidies			0.78	0.33	0	1	0.77	0.32	0	1
Subsidies are Publicly Shown in the Village or Township	Median	See (C) below.	0.46	0.42	0	1	0.47	0.41	0	1
Most Households Receive Subsidies in Cash			0.20	0.40	0	1	0.21	0.39	0	1
Most Households Receive Subsidies by Smart Card			0.78	0.41	0	1	0.71	0.43	0	1

Table 3. *Cont.*

Constructed Village Characteristics	Statistic Used[+]	Interpretation	Village Sample (n = 92)				Township Sample (n = 93)			
			Mean	SD	Min	Max	Mean	SD	Min	Max
Other Village/Township Characteristics (X^C)										
Village Has Newly Developed Wasteland on Sloped Land	Median	(C) Discrete indicator of village institution, based on the median response among village households in the sample.	0.07	0.25	0	1	0.06	0.25	0	1
Village Has Extended Cropland into Forest Area			0.07	0.25	0	1	0.08	0.27	0	1
Village Has Fallow Land			0.34	0.48	0	1	0.32	0.47	0	1
"A Lot" of HHs Participate in Rural Health Insurance			0.92	0.27	0	1	0.92	0.27	0	1
Average Share of Household Labor that is Migrant			0.37	0.21	0	0.87	0.35	0.21	0	0.87
Share of households in the village ...										
... with CCFP land that is formally registered agricultural land	Mean		0.81	0.30	0	1	0.80	0.30	0	1
... with a forest certificate		See (B) above.	0.61	0.44	0	1	0.61	0.44	0	1
... that have participated in collective reforms			0.48	0.42	0	1	0.49	0.41	0	1
... with collectively managed forestland			0.07	0.19	0	1	0.07	0.17	0	1
... that worked in forestry before CCFP			0.21	0.30	0	1	0.22	0.29	0	1
... that are ethnic minorities			0.17	0.35	0	1	0.16	0.34	0	1
... utilizing alternate fuels			0.28	0.34	0	1	0.27	0.32	0	1
... with sloped cropland			0.42	0.37	0	1	0.42	0.37	0	1

[+] Statistics calculated across the household sample for a given village; [++] these are indicated as "Forest Management" and "Afforestation" responsibilities; [+++] This can be either wasteland afforestation, or "closed-mountain" afforestation.

Table 4. Interval regression model of tree survivorship on household conversion of enrolled land form the cropland to forests program.

Variable	Only Household-Level (n = 2635)	Conditioning on Local Characteristics, Calculated as Averages at						
		...Village-Level (n = 1675)				...Township-Level (n = 1785)		
	(1)	(2)	(3)	(4)	(5)	(6)	(7)	(8)
Constant	70.050 ***	64.492 ***	66.240 ***	72.571 ***	68.076 ***	67.715 ***	81.890 ***	81.837 ***
		Regional Indicators						
Northwestern Province	−3.535	−3.780	−1.221	−0.228	0.868	−1.448	−1.278	−0.566
Southwestern Province	−4.390 **	−3.457 *	0.059	−0.094	−0.313	0.308	0.586	−0.204
Southeastern Province	−0.631	1.116	2.132	4.310	3.750	2.473	4.507	3.139
		Household Socioeconomic Characteristics						
Household Labor Population (>15 Years Old)	0.530	0.408	0.196	0.001	0.002	0.144	−0.067	0.005
Respondent Has High School or Above Education (1 = Yes)	0.017	−0.376	0.940	0.873	0.945	1.332	0.995	1.320
Respondent Has No Education or "Other" (1 = Yes)	−0.468	0.489	0.028	0.084	0.685	0.614	0.608	0.883
Share of Household Labor that is "Migrant"	−0.013	−0.015	−0.022	0.018	0.018	−0.017	0.018	0.016
Household Per Capita Crop Area Pre–CCFP (ha)	−3.417	−4.099	−8.065	−6.461	−6.251	−7.127	−5.401	−5.824
Household Per Capita Forest Area Pre–CCFP (ha)	1.676 ***	2.082 ***	2.603 **	2.873 ***	2.512 **	2.247 ***	2.233 ***	2.264 ***
Agriculture Is Household's Main Income Source (1 = Yes)	−0.656	−0.526	−0.777	−0.091	0.048	−0.579	0.165	−0.162
Household Is Poor in the Village (1 = Yes)	−2.314 *	−1.440	−1.336	0.006	−0.165	−1.024	−0.572	−1.007
Household Is Rich in the Village (1 = Yes)	2.037	1.257	0.812	0.944	1.301	−0.113	−0.134	−0.088
Household Members Worked Before in Afforestation (1 = Yes)	14.261	7.780	12.955	38.815 **	40.122 **	12.676	27.401 *	24.070 *
Household Has Cropland >25 Degrees (1 = Yes)	3.783 ***	3.023 **	3.812 **	−0.430	−0.424	4.097 **	0.171	0.011
Ethnic Minority? (1 = Yes)	−7.414 ***	−7.400 ***	−8.518 ***	−0.809	−0.880	−8.174 ***	−2.271	−3.038

Effect on Survival Rate (%)+

Table 4. *Cont.*

Variable	Only Household-Level (n = 2635)		Effect on Survival Rate (%) [+] Conditioning on Local Characteristics, Calculated as Averages at					
			...Village-Level (n = 1675)			...Township-Level (n = 1785)		
	(1)	(2)	(3)	(4)	(5)	(6)	(7)	(8)
Household-Level: Program Implementation								
Years in CCFP	4.948 **	3.311	6.371 **	7.581 ***	8.393 ***	5.599 **	4.455 **	3.996
× Worked Before in Afforestation	−4.871	−2.887	−4.457	−11.347 **	−12.056 **	−4.468	−8.078 *	−7.529 *
(Years in CCFP)^2	−0.386 **	−0.260 *	−0.605 ***	−0.697 ***	−0.692 ***	−0.541 ***	−0.454 **	−0.367 *
× Worked Before in Afforestation	0.492 *	0.347	0.466	0.848 ***	0.915 ***	0.464	0.638 **	0.628 **
Household CCFP Land Has Been Inspected Before (1 = Yes)		0.500	−8.483	−11.723 *		−9.685	−13.975 **	
× Years in CCFP		0.536	1.972	1.802 *		2.125	2.259 **	
Household CCFP Land is formally registered agricultural land		8.757 *	4.754	6.941		5.423	9.893 *	
× Years in CCFP		−0.875	−0.481	−0.702		−0.578	−1.178	
Household Has Other CCFP Afforestation Responsibilities		−1.429	−1.271	−0.590		−1.899	−1.287	
Household "Probably/Partially Understands" the Policy		−1.563	−0.988	−1.693 *	−2.094 **	−0.776	−1.726 *	−2.385 **
Household "Does not Understand" the Policy		−4.322 *	−5.575 *	−4.825 **	−5.416 ***	−4.726 *	−3.928 **	−4.588 **
Household "Knows" Their Responsibilities under CCFP		0.781	−3.586	1.801		−2.649	2.828 *	
Household Receives Subsidies in Cash		7.342 *	6.642	−0.565		5.277	1.210	
Household Receives Subsidies via Smart Card		6.842 **	6.031	7.366 *		4.833	7.766 **	
CCFP Land Management Type Is "Large HH" or "Large HH and Company"		−1.929	−5.675	−6.052		−6.028	−6.235 *	
Orchard Trees (1 = Yes)	−1.395	−1.853	−2.804	−2.803	−2.615	−2.539	−1.962	−1.794
Joint Type of Tree (e.g., Bamboo) (1 = Yes)	1.949	2.031	−1.886	−2.115	−2.377	−1.215	−1.540	−1.847

Table 4. *Cont.*

Shrubs (1 = Yes)	1.441	0.845	0.535	3.075	2.942	0.985	3.411	3.291
Had to change tree types on CCFP land? (1 = Yes)	-7.687	-6.147	1.125	-0.556	-2.954	0.848	-2.718	-6.313
x Years in CCFP	0.269	0.177	-0.251	-0.187	0.068	-0.240	0.048	0.458
CCFP Land Main Type Is Sloped (1 = Yes)	0.780	-0.876	-1.454	-1.998	-2.418	-1.557	-1.557	-1.842
CCFP Land Main Type Is Desertified (1 = Yes)	-1.786	-1.349	0.308	0.096	0.337	0.694	2.052	2.459
CCFP Plots Average Distance from Home (kilometers)	-1.278 ***	-1.206 ***	-1.075 ***	-0.989 ***	-1.071 ***	-1.093 ***	-0.986 ***	-1.084 ***
CCFP Land was Hit by Disaster (1 = Yes)	-2.500	-2.056	0.521	-0.148	-0.405	-0.190	-1.233	-1.412
Intercropping on CCFP Land (1 = Yes)	-0.734	-0.345	-2.926	-0.945	-0.957	-2.233	-0.535	-0.651
HH Per Capita CCFP Land Area (ha)	7.486 ***	5.590 **	9.990	10.240	9.733	7.420	6.039	5.642
Average Yield of CCFP Land was High, Pre–CCFP	-2.162	-3.011	-3.207	-1.047	-1.240	-2.960	-0.694	-1.152
Household-Level: Other Policies with Potential Impacts								
HH Forest Land Is Collectively Managed	5.581 ***	6.073 **	4.079 **	2.468	6.229 ***		5.194 **	3.841 *
Received Forest Certification (as part of collective forest sector reforms)	1.841	1.645	2.366	2.438	2.049		2.731	2.551
Local Economic and Institutional Characteristics (Village/Township Averages) [++]								
Share of Households with CCFP Land Inspected				0.155 ***	0.195 ***		0.135 **	0.174 ***
Share of Household CCFP Land "In the Books"				-0.007	0.012		0.009	0.020
Share of Households with Intercropping on CCFP Land				-0.061	-0.056		-0.063	-0.065
Share of Households That "Understand" the Policy				-0.013	-0.022		-0.022	-0.031
Share of Households That "Do not Understand" the Policy				-0.077	-0.076		-0.074	-0.061
Share of Households That "Know Their Responsibilities" under CCFP				-0.186 ***	-0.141 **		-0.209 ***	-0.137 **
Share of Households with Other CCFP Afforestation Responsibilities				-0.025	-0.030		-0.013	-0.024
Subsidies Are Publically Shown in the Village or Township				0.007	0.006		0.022	0.022
Share of Households That Received Full Subsidies				-0.079 **	-0.074 *		-0.093 ***	-0.082 **
Most Households Receive Subsidies in Cash				0.104 **	0.045		0.087 *	0.047
Number of Years Village or Township has Been Implementing CCFP				0.554	0.437		0.205	0.051

Table 4. *Cont.*

Village/Township Has Newly Developed Wasteland on Sloped Land (1 = Yes) +++			4.443 *	4.338		6.140 **	6.048 **
Village/Township Has Extended Cropland in Forest Area (1 = Yes) +++			1.154	1.347		0.437	1.020
Village Has Fallow Land (1 = Yes) +++			−2.355	−2.229		−4.022 *	−3.958 *
"A Lot" of Households in the Village Participate in Rural Health Insurance (1 = Yes) +++			−4.526	−4.674		−1.873	−1.688
Share of Households with Forest Certificate			−0.014	0.002		−0.012	0.010
Share of Households that Have Participated in Collective Forest Sector Reforms			−0.005	−0.009		−0.014	−0.017
Share of Households that are Ethnic Minorities			−0.123 ***	−0.107 ***		−0.101 ***	−0.074 **
Share of Households that Worked in Forestry Before CCFP			0.016	0.018		0.006	0.011
Share of Households Utilizing Alternate Fuels			0.014	0.005		−0.005	−0.013
Share of Households with Sloped Cropland			0.088 **	0.096 ***		0.093 ***	0.102 ***
Share of Households with Collectively Managed Forestland			0.059	0.074		0.059	0.071
Average Share of Household Labor that is Migrant			−0.001 *	−0.001 *		−0.001	−0.001
Log-Likelihood Function	−3981.81	−2322.40	−2204.11	−2230.99	−2500.38	−2383.05	−2421.22
McFadden's Adjusted R^2	0.039	0.078	0.115	0.108	0.072	0.106	0.096

… = Share of households in the village; * Significant at 10%/ ** Significant at 5%/ *** Significant at 1%; + Coefficients can be interpreted as the increase in survival rate (%) for an increase in one unit of the variable in question; ++ Village and township averages were calculated for villages/townships with data from at least 15 households per village or township; +++ Calculated as a discrete Yes or No based on the average response of households in the village or township. Models utilize robust standard errors clustered at the village.

All models are estimated with robust standard errors clustered at the village level. Since the interval regression explains the underlying (unobserved) survival rate, parameter estimates are interpreted as the marginal effect of a one-unit increase of the variable in question on survival rate (e.g., a parameter estimate of 10 means that a one unit increase of the variable in question is associated with an increase in the survival rate by 10 percentage points). All household and village-share variables are scaled to a 1 to 100 range so that parameter estimates for these can be interpreted directly, as the impact of an increase in village share by 1% on survival rate.

Overall, model results broadly accord with expectations. In terms of household characteristics, households with higher levels of human capital in forestry activities appear to be better at keeping program trees alive, as would be expected; having higher per capita forest area, and a member who has worked before in afforestation, pre-CCFP, are both positively associated with survival rate in all specifications, with the former being statistically significant in all. Landholders with experience in forest management and tree planting are more likely to enthusiastically adopt re/afforestation activities than those who do not have such experience. Other work in the literature finds that poor incorporation of forest management knowledge transfer, resulting in a lack of sufficient tending and thinning, is associated with low plantation survival rates in China, while CCFP outcomes in locations that that are historically not forest areas—and therefore where local forest management experience is low—have been unsatisfactory [29]. Lack of technical support has often been noted as a key shortcoming in program implementation; such support could help reduce perception of risk and improve outcomes [30].

Though less robust, results also provide weak evidence that household capital constraints adversely affect ability to manage program-planted trees; being "rich" in the village (and therefore likely having more assets to contribute to production activities) is in general positively associated with survival rate, while being "poor" is negatively associated.

Reflecting the time costs of labor, each additional kilometer that CCFP plots are from the household's home on average is associated with a decrease in survival rates by between 0.99% and 1.28%, with parameter estimates statistically significant across all specifications. Smallholders empowered to select plots for re/afforestation can often choose plots located further from their homes or roads to reduce the opportunity costs (in terms of income and food security) of retiring more accessible agricultural land [30]. Indeed, Xu and colleagues [10] find that when households have autonomy in whether or not to participate in CCFP, they strongly tend to choose plots farther away from the homestead. However, this comes at the cost of environmental outcomes. Bennett and colleagues [24] found that when CCFP participants are allowed to select plots for conversion, famers choose less fertile and more remote plots, with this resulting in lower survival rates [24].

Capturing labor opportunity costs, higher village or township average share of participant household labor engaged in off-farm migratory work is associated with lower survival rates at the household level, with this statistically significant in the models that include village characteristics. These results suggest that in locales where regional off-farm opportunities are more plentiful, households tend to redistribute labor effort towards these more profitable activities and away from tending program-planted trees.

In terms of the effects of local characteristics, relative endowments of land *versus* labor at the local level produce statistically significant results that accord with expectations, whereby having higher labor endowments per unit of land is broadly associated with higher survival rates at the household level (*i.e.*, ostensibly due to higher availability of labor to invest in managing program-planted trees). For example, being in a village with fallow land—suggesting a higher land/labor ratio (though possibly also reflecting a local economy where off-farm work is more plentiful)—is associated with lower survivorship at the household level. Similarly, being in a village with newly-developed wasteland on sloping land and where a larger share of sample households still cultivates crops on sloping land—both of which reflect lower land/labor ratios, which create incentives to expand cultivated area—are associated with higher survival rates on household land.

Model results also suggest that learning-by-doing and specialization improves program outcomes. Specifically, "Years in CCFP" and "(Years in CCFP)^2" are both statistically significant in most specifications, with their relative signs suggesting that households improve their ability to manage trees the longer they are in the program, with this effect decreasing over time. The interactions of these variables with whether or not a household member has worked before in afforestation also indicates that households with forestry experience start out with an advantage in managing forest area, though with this decreasing over time (*i.e.*, for household with members that have worked before in forestry, the base survival begins at 39% in the village models, with this decreasing but leveling out over time).

Management structure on household enrolled land clearly appears to influence survival rates. Higher household per capita area enrolled in CCFP is positively associated with survival rates, suggesting that household specialization in providing program-targeted ecological services improves survival rates. Conversely, if household land is managed by a contracted third party under the "Large Household" or "Large Household + Company" management types—which generally consist of a large portion or all of village CCFP land managed by a few large households or a combination of this and a contracted outside implementation unit—survival rates tend to be lower, with this effect statistically significant in one of five specifications.

Regarding ecological factors, orchard crops appear to have lower survival rates in general in comparison to timber trees (the omitted category), while shrubby tree crops have better survival rates. Planting program trees primarily on sloping land, and household CCFP land being affected by disaster (*i.e.*, drought, flood or pest/diseases), are both associated with reduced survivorship, though these effects are not robust across specifications. Having CCFP land that is formally registered as agricultural land is positively associated with survival rate, and is statistically significant in two of five specifications. This likely captures the effect of land quality on survival rates, since formally registered agricultural land is generally of better quality (from conversations with SFA officials and observations made in the field).

Also, households who had to change the tree type initially planted on enrolled land due to low survival rates tend to have lower survival rates for currently planted trees, even when controlling for the number of years in the program (*i.e.*, via interacting this with number of years in the program). This suggests that poor initial selection of tree species that are adapted to local conditions can delay achievement of longer-term outcomes.

Finally, of relevance for the program's ecological and economic sustainability, intercropping on enrolled land is generally associated with lower survival rates. The degree to which households are allowed to intercrop on enrolled land has specific relevance for household livelihoods, especially in land-scarce regions with low off-farm wage labor opportunities. Note, however, that this effect is similar at the village level, with share of participant households at the village-level that are intercropping also negatively associated with survival rate at the household level. This suggests that this variable could also be picking up local institutional effects, whereby incentives to achieve higher survival rates are weaker in locales with a less rigorous and more permissive implementation regimes.

In terms of institutional impacts, several variables appear to be strongly associated with survival rates. First, the fact that a household "does not understand" the CCFP is negatively associated with survival rates, with this statistically significant across all specifications. More intriguing, the share of households at either the village or township level that "do not understand" the CCFP policy is also negatively associated with survival rates at the household level. This suggests that the degree to which local governments systematically consult and engage with participant households and communities during program implementation has an important effect on program outcomes.

Participant households having additional afforestation responsibilities apart from afforesting cropland—which include either afforestation of "wasteland" or "closed-mountain" afforestation—tend to have lower survival rates, as do households in villages with a higher share of participant households that have these additional responsibilities. This suggests that additional program demands on household labor as part of program participation comes at the potential cost of weakening outcomes.

Finally, variables capturing the inspection regime reveal a somewhat complex picture. At the household level, direct correlation between whether CCFP land has been inspected before and survival rate on CCFP land is positive and highly statistically significant in the data set. Indeed, inspection is positive but statistically insignificant in the household-level model. However, in specifications using village (and township) characteristics, it becomes negative and statistically significant in two of five specifications, while share of sample participant households in the village (or township) whose CCFP land has been inspected is very positively associated with household-level survivorship, statistically significant in all specifications where it is included. Due to concerns about the direction of causality, the household-level inspection variable was also interacted with years in the program, which finds that each additional year that inspected CCFP land has been in the program is positively associated with survival rate, with this statistically significant in three of five specifications. Overall, this suggests that while a strong and well-functioning inspection regime at the village level is strongly associated with better survivorship, household-level effects are more complicated, likely capturing a range of different factors including informational asymmetries and complex behavioral dynamics not fully captured by the other explanatory variables in the model.

That some variables likely vary systematically by region could be potentially complicating identification of their effects, since these could be confounded with other unobserved regional impacts, including eco-regions, biome and local climate. The lack of detailed data on tree species in the survey, for example, means that each tree category likely encompasses significant heterogeneity, with the nature of this varying regionally (for example the SFA documents 72 types of ecological trees, 21 economic trees, 51 shrubs and 20 joint trees in southern China; and 42 type of ecological trees, 16 economic trees, 53 shrubs and 16 joint trees in northern China [31]). As such, to disentangle these effects and improve identification, an additional set of specifications were explored wherein a subset of variables were interacted with the regional indicators. Variables deemed likely to be confounded with regional effects were as follows: whether or not the household has sloped cropland; household per capita CCFP land area; tree type; village share of participant households that are intercropping on their CCFP land; village share of participant households that have other afforestation responsibilities as part of CCFP; whether or not the village has fallow land; village share of households that are ethnic minorities; and village average of household share of labor in off-farm migrant work.

Model results are presented in Tables 5 and 6 below. Interactions with the regional indicators indeed appear to disentangle some of these effects and improve overall model performance, with (11) having the best measure of fit of any of the specifications. The associated survival rate impacts of tree type indeed vary in important ways across regions. "joint type" of trees interacted with "southeast" in a positive and significant way, with large parameter estimates in all specifications, likely picking up the faster-growing bamboo plantations of the southeastern provinces ("joint type" covers tree types which overlap both "ecological" and "economic" delineations, as well as bamboo, which is considered as a special tree type in China's forestry statistics system [31]). Similarly, "shrubs" in the southwest are also associated with a large positive boost to survival rates, significant in all specifications.

Village-level characteristics all vary regionally in important ways: the impacts of village share of participant households that intercrop on CCFP land; whether or not the village has fallow land; the village share of participant households that are ethnic minorities; and the average village share of participant household labor that is migrant. The main effect of the village share of households intercropping on their CCFP land is large, negative and significant, again suggesting that a village-level implementation regime that is more permissive regarding how households utilize CCFP land might result in reduced effectiveness in achieving targeted survival rate targets. However, this appears to be offset by significant regional impacts for the northwest and southwest. Whether the village has fallow land also retains its negative impact in the main effect, but for the southwest has a net positive impact on survival rates, suggesting that factors other than village land/labor ratio could be driving this relationship.

The strong and statistically significant negative effect of village share of participant households that are ethnic minorities disappears when the village variable is interacted with regional indicators. This confirms the suspicion that this variable could be picking up important unobserved regional characteristics that have a bearing on the effectiveness of program implementation, such as remoteness (e.g., from regional government seat, or regional markets) or distinct differences in local customary governance structures for forestry and agriculture. In particular, interactions between the village share of participant households that are ethnic minorities and the regional indicators of southwest and northwest produce highly significant and negative parameter estimates, while the main effect becomes positive and statistically insignificant, and whether the household is an ethnic minority also becomes statistically insignificant from being previously highly statistically significant with negative impacts in the earlier models.

Interestingly, the regional interactions with the ethnic minority share could also be capturing important social dimensions or indigenous knowledge. It has been found (unsurprisingly) that a farmer's ethnic group influences perceptions regarding tree planting [32]. Approaches to silviculture also can vary among ethnic groups based on the length of time in a given area and on experience planting specific species or combinations thereof. In southwestern China, different communities employ different rotation and fallow periods, as well as intercropping of tree species to avoid declines in productivity that have been associated with monocultures employed in state forestry programs and by ethnic Han groups with a shorter history in the area [30].

Table 5. Interval regression model of tree survivorship on household conversion of cropland to forests program (CCFP) enrolled land with regional interactions.

Variable	Direct Effect (%) +	Household Only (9) Interactions with regional indicators		
		Northwest +	Southwest +	Southeast +
Constant/Regional Dummies	64.668 ***	-2.699	-6.499 **	-2.285
Household Socioeconomic Characteristics				
Household Labor Population (>15 Years Old)	0.532			
Respondent Has High school or Above Education (1 = Yes)	-0.479			
Respondent Has No Education or "Other" (1 = Yes)	0.817			
Share of Household Labor That is "Migrant"	-0.013			
HH PC Crop Area Pre-CCFP (ha)	-2.26			
HH PC Forest Area Pre-CCFP (ha)	2.318 ***			
Agriculture Is HH's Main Income Source (1 = Yes)	-0.032			
HH Is Poor in the Village (1 = Yes)	-1.2			
HH Is Rich in the Village (1 = Yes)	1.38			
HH Members Worked Before in Afforestation (1 = Yes)	8.102			
HH Has cropland > 25 Degrees (1 = Yes)	3.351 **	2.09	-0.765	-3.448
Ethnic Minority? (1 = Yes)	-7.583 ***			
Household-Level: Program Implementation				
Years in CCFP	3.708 *			
× Worked Before in Afforestation	-2.87			
(Years in CCFP)²	-0.298 *			
× Worked before in Afforestation	0.332			
HH CCFP Land Has Been Inspected Before (1 = Yes)	-0.734			
× Years in CCFP	0.688			
Household CCFP Land Is "In the Books"	7.284			
× Years in CCFP	-0.65			

Table 5. *Cont.*

Variable	Direct Effect (%)[+]	Household Only (9) Interactions with regional indicators		
		Northwest [+]	Southwest [+]	Southeast [+]
Household-Level: Program Implementation				
HH Has Other CCFP Afforestation Responsibilities	−1.849			
Household "Probably/Partially Understands" the Policy	−2.004 *			
Household "Does not Understand" the Policy	−4.382 **			
Household "Knows" Their Responsibilities under CCFP	0.672			
Household Receives Subsidies in Cash	6.815 *			
Household Receives Subsidies via Smart Card	6.535 **			
CCFP Land Management Type Is "Large HH" or "Large HH & Company"	−2.639			
Orchard Trees (1 = Yes)	−4.481 **	−0.448	5.376	8.948 *
Joint Type of Tree (e.g., Bamboo) (1 = Yes)	−5.948	5.603	13.025 ***	14.976 ***
Shrubs (1 = Yes)	−6.077	2.061	15.595 ***	11.668 **
Had to Change Trees Types on CCFP Land? (1 = Yes)	−7.021			
× Years in CCFP	0.335			
CCFP Land Main Type Is Sloped (1 = Yes)	−0.477			
CCFP Land Main Type Is Desertified (1 = Yes)	−1.303			
CCFP Plots Average Distance from Home	−1.17 ***			
CCFP Land Was Hit by Disaster (1 = Yes)	−2.017			
Intercropping on CCFP Land (1 = Yes)	−0.162			
HH Per Capita CCFP Land Area (ha)	8.103 **	−9.532	−8.3	0.509
Average Yield of CCFP Land Was High, Pre-CCFP	−3.027			
Household-Level: Other Policies with Potential Impacts				
HH Forest Land is Collectively Managed	5.346 ***			
Received Forest Certification (as part of collective forest sector reforms)	1.764			
Log-Likelihood Function	−3866.428			
McFadden's Adjusted *R²*	0.058			

* Significant a 10%/** Significant at 5%/*** Significant at 1%; [+] Coefficients can be interpreted as the increase in survival rate (%) for an increase in one unit of the variable in question; interaction term coefficients should be added to the direct effect to calculate the total effect for each regional subsample; Models utilize robust standard errors clustered at the village.

Table 6. Interval regression model of tree survivorship on household conversion of cropland to forests program (CCFP) enrolled land with regional interactions.

Variable	Local Variables at the Village-Level							
	(10)				(11)			
	Direct	Interactions with regional indicators			Direct	Interactions with regional indicators		
	Effect (%) [+]	Northwest [+]	Southwest [+]	Southeast [+]	Effect (%) [+]	Northwest [+]	Southwest [+]	Southeast [+]
Constant/Regional Dummies	57.828 ***	20.359 ***	8.468 *	−11.161	63.096 ***	23.579 ***	13.315 **	−10.283
Household Socioeconomic Characteristics								
Household Labor Population (>15 Years Old)	0.115				0.189			
Respondent Has High school or Above Education (1 = Yes)	0.146				0.364			
Respondent Has No Education or "Other" (1 = Yes)	−0.013				−0.380			
Share of Household Labor that Is "Migrant"	0.018				0.016			
HH PC Crop Area Pre-CCFP (ha)	−2.520				−4.442			
HH PC Forest Area Pre-CCFP (ha)	2.051 *				2.246 *			
Agriculture is HH's Main Income Source (1 = Yes)	−0.451				−0.311			
HH Is Poor in the Village (1 = Yes)	−0.323				−0.777			
HH Is Rich in the Village (1 = Yes)	1.256				0.979			
HH Members Worked Before in Afforestation (1 = Yes)	46.610 ***				36.636 **			
HH Has Cropland > 25 Degrees (1 = Yes)	1.669	−2.624	−3.631	−4.329	2.113	−2.963	−4.965	−3.781
Ethnic Minority? (1 = Yes)	−0.731				−0.730			
Household-Level: Program Implementation								
Years in CCFP	7.582 ***				6.477 **			
× Worked Before in Afforestation	−13.630 ***				−10.577 **			
(Years in CCFP) [2]	−0.690 ***				−0.583 ***			
× Worked Before in Afforestation	0.974 ***				0.754 ***			
HH CCFP Land Has Been Inspected Before (1 = Yes)	−8.653				−8.267			

Table 6. Cont.

Variable	Local Variables at the Village-Level							
	(10)				(11)			
	Direct Effect (%) [+]	Interactions with regional indicators			Direct Effect (%) [+]	Interactions with regional indicators		
		Northwest [+]	Southwest [+]	Southeast [+]		Northwest [+]	Southwest [+]	Southeast [+]
× Years in CCFP	1.263				1.205			
Household CCFP Land is Registered Agric. Land	4.287				7.776			
× Years in CCFP	−0.336				−0.817			
HH Has Other CCFP Afforestation Responsibilities	−0.552				−0.396			
Household "Probably/Partially Understands" the Policy	−1.634				−1.470			
Household "Does not Understand" the Policy	−4.975 ***				−4.708 **			
Household "Knows" Their Responsibilities under CCFP	2.030				2.081			
Household Receives Subsidies in Cash	1.176				1.451			
Household Receives Subsidies via Smart Card	7.274 **				6.699 *			
CCFP Land Management Type is "Large HH" or "Large HH and Company"	−5.564 *				−5.204			
Orchard Trees (1 = Yes)	0.842	−11.689 **	−5.638	4.168	3.274	−16.046 ***	−7.391 *	0.541
Joint Type of Tree (e.g., Bamboo) (1 = Yes)	−6.061 *	1.730	1.968	18.134 **	−3.078	−4.785	−0.667	13.458 **
Shrubs (1 = Yes)	−11.697 **	7.667	16.181 ***	7.605	−13.266 **	5.213	20.284 ***	13.231
Had to change trees types on CCFP land? (1 = Yes)	−2.690				−0.744			
× Years in CCFP	−0.061				−0.274			
CCFP Land Main Type is Sloped (1 = Yes)	−1.674				−1.578			
CCFP Land Main Type is Desertified (1 = Yes)	0.249				0.238			
CCFP Plots Average Distance from Home	−0.539 *				−0.677 **			
CCFP Land was Hit By Disaster (1 = Yes)	0.455				0.203			
Intercropping on CCFP Land (1 = Yes)	−0.583				−0.450			
HH Per Capita CCFP Land Area (ha)	9.602	−3.117	−8.724	−10.167	7.874	0.767	−2.055	−2.668
Average Yield of CCFP Land was High, Pre-CCFP	−0.786				−0.149			

Table 6. *Cont.*

Variable	Direct Effect (%) [+]	(10) Interactions with regional indicators			Direct Effect (%) [+]	(11) Interactions with regional indicators		
		Northwest [+]	Southwest [+]	Southeast [+]		Northwest [+]	Southwest [+]	Southeast [+]
Household-Level: Other Policies with Potential Impacts								
HH Forest Land is Collectively Managed	4.544 **				4.486 **			
Received Forest Certification	2.370				2.081			
Share of Households with CCFP Land Inspected	0.187 ***				0.206 ***			
Share of Household CCFP Land "In the Books"	0.005				0.013			
Share of Households with Intercropping on CCFP Land	−0.131 **				−0.288 ***	0.319 ***	0.250 **	0.424 ***
Share of Households That "Understand" the Policy	−0.003				0.039			
Share of Households That "Don't Understand" the Policy	−0.002				−0.016			
Share of Households That "Know Their Responsibilities" under CCFP	−0.148 ***				−0.146 ***			
Share of Households with Other CCFP Afforestation Responsibilities	−0.045				0.177 **	−0.254 ***	−0.294 ***	−0.168 ***
Subsidies Are Publically Shown in the Village or Township	0.027				0.011			
Share of Households that Received Full Subsidies	−0.086 **				−0.085 ***			
Share of Households that Received Subsidies in Cash	0.086 **				0.086 *			
Number of Years Village Has Been Implementing CCFP	0.475				0.109			
Village Has Newly Developed Wasteland on Sloped Land (1 = Yes) [+++]	2.325				0.012			
Village Has Extended Cropland in Forest Area (1 = Yes) [+++]	−1.897				0.052			
Village Has Fallow Land (1 = Yes) [+++]	−2.146	−10.988	6.172	−12.990 **	−7.191 *	−3.915	10.528 **	−7.581
"A Lot" of HHs in the Village Enrolled in Rural Health Insurance [+++]	−4.950				−8.459 *			
Share of Households with Forest Certificate	−0.003				0.015			

Table 6. *Cont.*

	Local Variables at the Village-Level							
	(10)				(11)			
	Direct	Interactions with regional indicators			Direct	Interactions with regional indicators		
Variable	Effect (%) +	Northwest +	Southwest +	Southeast +	Effect (%) +	Northwest +	Southwest +	Southeast +
Household-Level: Other Policies with Potential Impacts								
Share of Households That Have Participated in Collective Reform	−0.019				−0.040			
Share of Households That Are Ethnic Minorities	−0.002	−0.267 **	−0.123 *	0.072	0.001	−0.190 *	−0.141 **	−0.098
Share of Households That Worked in Forestry Before CCFP	0.030				−0.031			
Share of Households Utilizing Alternate Fuels	0.032				0.052			
Share of Households with Sloped Cropland	0.095 ***				0.081 ***			
Share of Households with Collectively Managed Forestland	−0.008				−0.022			
Average Share of Household Labor That Is Migrant	0.000	−0.002	−0.002	0.004	0.001	−0.003	−0.003	0.002
Log-Likelihood Function	−2105.72				−2071.63			
McFadden's Adjusted R^2	0.138				0.146			

* Significant a 10%/** Significant at 5%/*** Significant at 1%; + Coefficients can be interpreted as the increase in survival rate (%) for an increase in one unit of the variable in question. Interaction term coefficients should be added to the direct effect to calculate the total effect for each regional subsample; ++ Village and township averages were calculated for villages/townships with data from at least 15 households per village or township; +++ Calculated as a discrete Yes or No based on the average response of households in the village or township. Models utilize robust standard errors clustered at the village.

5. Conclusions

The CCFP is representative of China's ongoing transition from an extractive economic growth model to one that is more environmentally sustainable. This accords with international trends, which have seen a net increase in forest cover occurring in several Asian countries in recent decades—most notably in China, Vietnam, Bhutan and India —in what has been termed a "forest transition" [33–36]. While urbanization and de-agrarianization have been identified as major causes of forest transition elsewhere (and particularly in developed/western countries), in Asia, government programs promoting timber production and provision of ecosystems services have been a major driver of the increase in land designated as forest [34,37].

It is important to improve the understanding of which program design elements are most effective for ensuring that such programs successfully achieve their re/afforestation outcomes, and are able to sustainably facilitate such a transition in Asia and elsewhere. This is especially important in the presence of heterogeneity in local institutional and socioeconomic conditions. As the world's largest re/afforestation Payment for Ecosystem Services program, encompassing a wide array of ecological and socioeconomic conditions, the CCFP provides an excellent opportunity to do this.

In general, our analysis provides evidence that household and local socioeconomic characteristics, as well as local program implementation regime, all play important roles in determining outcomes. At the household level, households with pre-existing training and experience in forestry, as well as higher labor endowments relative to land, do better at managing trees. Time in the program also appears to increase tree survivorship, suggesting that important learning-by-doing effects are taking place as well. Conversely, higher opportunity costs for either land or labor have the opposite effect. Related to this, households in areas with relatively abundant labor (*i.e.*, high labor/land ratios), and with relatively poor access to off-farm work opportunities, in general do better at keeping their trees alive, likely due to having more labor to invest in planting and management. This suggests that providing ongoing technical support and training to households to help them improve their forestry skills could have important knock-on effects on program ecological outcomes. These findings accord with prior findings of Bennett and colleagues [24].

Our findings also provide some strong evidence that the local implementation regime has an important effect on outcomes, and that both incentives and monitoring are critical. First, the most important of these results is that the degree to which program managers have consulted with participant communities and households has a strong positive influence on outcomes. This is captured in the degree to which households indicate that they do not understand the program, and for those who are in villages where a large share of participant households also indicate that they do not understand the program, both which result in households achieving lower survival rate outcomes. In general, this finding resonates with accepted best practice for Payments for Ecosystem Services programs, which stresses that ongoing consultation with communities during all stages of program development and implementation will help to both improve outcomes and reduce costs.

Another important result is that in places where intercropping on enrolled land is more predominant, households are less successful at keeping program trees alive. While more work needs to be done to understand the specifics of this relationship, these results suggest that tradeoffs exist between how rural livelihoods issues are addressed and program environmental goals; allowing households to intercrop on CCFP enrolled land might help to minimize impacts on food and cash crop production, but at the expense of tree survivorship. It is also possible that these results reflect the technical challenges evident in engendering effective and sustainable agroforestry regimes. Viewed from a different angle, however, it is also possible that this could be capturing the adverse impacts on survivorship of more permissive, less rigorous implementation regimes.

Finally, a number of intriguing results suggest that improvements could be made to the program's subsidy and inspection regime. In particular, the village share of households with CCFP land that has been inspected has a strong positive relationship on household-level tree survivorship, while whether or not a household's CCFP land has been inspected has a strong negative one. This suggests that a well-managed local inspection regime clearly does better at achieving outcomes. However, ongoing monitoring and inspections might be required to ensure that program goals are sustainably met, possibly due to the fact post-inspection, households face weaker incentives to continue maintaining survival rates. Similarly, the strong negative relationship between household-level survivorship and the share of households in a village that have received full subsidies also suggests that once the program winds down in a particular locale, thus reducing local government program implementation efforts, households will at the very least begin to curtail efforts at managing program-planted trees. This suggests that some degree of ongoing subsidy support and monitoring will be needed for CCFP forestry outcomes to be sustainably maintained and eventually consolidated, though the degree to which this is the case likely varies significantly by locale.

Admittedly, caution should be exercised when using the results of this analysis to make larger inferences regarding the CCFP's success in incentivizing household delivery of targeted forest ecosystem benefits, especially for off-site benefits. Forests provide a complex array of ecological services, with tree survivorship alone falling far short of capturing the spatial and temporal complexity of the underlying processes. Indeed, Le and colleagues argue that reforestation assessments should not be based on success indicators alone, but should incorporate the drivers of success, which encompass an array of biophysical, socioeconomic, institutional and project characteristics [22]. Ongoing monitoring and evaluation work for the CCFP should thus broaden its portfolio of indicators, especially regarding targeted ecological outcomes. This is particularly important given observations that China's afforestation statistics could be hiding a significant degree of exotic tree species planting, which could be having adverse impacts on biodiversity habitat [38].

The nature of the dependent variables used for this analysis, combined with the lack of detailed information on tree species planted and other important biophysical indicators (e.g., location within watersheds, soil type, microclimate, *etc.*), also limit the ability of this data set to capture on-site environmental outcomes. Lack of counterfactuals or detailed measures of *ex ante* household socioeconomic characteristics also limit the ability to rigorously identify program environmental impacts at the household level.

The results of this analysis are nonetheless revealing and valuable. They are based on what is arguably one of the largest and most representative samples yet available for evaluating the CCFP; other work has generally relied on much smaller, more regionally restricted survey data sets e.g., [10,11,15]. Model results, furthermore, suggest that real relationships are being uncovered;

several parameter estimates are found to be robust across specifications, statistically significant, and of the expected signs. Such findings accord with earlier findings by Bennett and colleagues [24] and suggest that with improved collection of a wider range of more detailed household and local-level indicators of targeted forest ecosystem services delivery, especially for the FEDRC's ongoing monitoring data, a wealth of additional insights into the impacts of program design elements on ecological outcomes could be obtained with relatively little additional effort. The sheer scale and range of local conditions encompassed by the CCFP, furthermore, suggest that such insights could prove a treasure trove for both domestic and international policymakers and practitioners.

Acknowledgments

The survey from which the data in this article was drawn was made possible by support from the CCFP Social Economic Monitoring Project, China National Forestry Economics and Development Research Center (FEDRC) of the State Forestry Administration of China. We would specifically like to acknowledge the team members of the FEDRC (Huang Dong, Yuan Mei, Peng Wei) and Beijing Forestry University (Wu Tao, Tian Seng Nan, Xu Ding, Zhang Cao, Su Yue Xiu) who were involved in the survey design and data collection. The writing of this article was supported by the Center for International Forestry Research (CIFOR) using funds from the UK Department for International Development (DFID) KNOWFOR Programme, which aims "to increase the value and impact of forest and tree-related knowledge by improving dissemination, strengthening knowledge pathways and increasing uptake by key forest sector stakeholders such as policymakers and practitioners." The authors would also like to gratefully acknowledge the efforts of the anonymous reviewers and editors of the special issue for their contribution towards improving the paper.

Author Contributions

Chen Xie and Daoli Peng designed the survey instrument, while Daoli Peng, Tao Wu and Dong Huang implemented the survey. Chen Xie cleaned and organized the final dataset. Michael T. Bennett formulated and implemented the econometric strategy, and wrote up the section on the modeling results. Chen Xie, Nicholas J. Hogarth, Louis Putzel and Michael T. Bennett all contributed to final write up and interpretation of model results and implications.

Conflicts of Interest

The authors declare no conflict of interest.

References

1. Bennett, M.T. China's Sloping Land Conversion Program: Institutional Innovation or Business as Usual? *Ecol. Econ.* **2008**, *65*, 699–711.
2. Hyde, W.F.; Belcher, B.M.; Xu, J. *China's Forests: Global Lessons from Market Reforms*; Resources for the Future: Washington, DC, USA, 2003.

3. Liu, C. *An Economic and Environmental Evaluation of the Natural Forest Protection Program*; National Forest Economics and Development Research Center (FEDRC), State Forestry Administration (SFA): Beijing, China, 2002.

4. Xu, J. The political, social, and ecological transformation of a landscape: The case of rubber in Xishuangbanna, China. *Mount. Res. Dev.* **2006**, *26*, 254–262.

5. China Council for International Cooperation on Environment and Development-Western China Forests and Grassland Task Force. In *Implementing the Natural Forest Protection Program and the Sloping Land Conversion Program: Lessons and Policy Recommendations*; Xu, J., Katsigris, E., White, T.A., Eds.; China Forestry Publishing House: Beijing, China, 2002.

6. Démurger, S.; Hou, Y.; Yang, W. *Forest Management Policies and Resource Balance in China: An Assessment of the Current Situation*; Groupe d'Analyse et de Théorie Economique (GATE), Ecole Normale Supérieure; Centre national de la recherche scientifique (CNRS): Lyon, France, 2007.

7. Démurger, S.; Hou, Y.; Yang, W. Forest management policies and resource balance in China: An assessment of the current situation. *J. Environ. Dev.* **2009**, *18*, 17–41.

8. SFA. *Forestry Development Annual Report*; China Forestry Publishing Press: Beijing, China, 2013.

9. Kelly, P.; Huo, X.X. Land Retirement and Nonfarm Labor Participation: An Analysis of China's Sloping Land Conversion Program. *World Dev.* **2013**, *48*, 156–169.

10. Xu, J.T.; Tao, R.; Xu, Z.G.; Bennett, M.T. China's Sloping Land Conversion Program: Does Expansion Equal Success? *Land Econ.* **2010**, *86*, 219–244.

11. Grosjean, P.; Kontoleon, A. How Sustainable and Sustainable Development Programs? The Case of the Sloping Land Conversion Program in China. *World Dev.* **2009**, *37*, 268–285.

12. Uchida, E.; Xu, J.; Xu, Z.; Rozelle, S. Are the poor benefiting from China's land conservation program? *Environ. Dev. Econ.* **2007**, *12*, 593–620.

13. Groom, B.; Grosjean, P.; Kontoleon, A.; Swanson, T.; Zhang, S. Relaxing constraints with compensation: Evaluating Off-farm labor responses to reforestation policy in China. In Proceedings of the Third World Congress of Environmental and Resource Economists, Kyoto, Japan, 2006.

14. Xie, C.; Zhao, Z.; Liang, D.; Zhang, L.; Dau, G.; Wang, X. Livelihood Impacts of the Conversion of Cropland to Forested Grassland Program. *J. Environ. Plann. Manag.* **2006**, *49*, 555–570.

15. Uchida, E.; Xu, J.T.; Rozelle, S. Grain for Green: Cost Effectiveness and Sustainability of China's Conservation Set-Aside Program. *Land Econ.* **2005**, *81*, 247–264.

16. Wang, X.H.; Bennett, J.; Xie, C.; Zhang, Z.T.; Dan, L. Estimating non-market environmental benefits of the conversion of cropland to forests and grassland program: A choice modeling approach. *Ecol. Econ.* **2006**, *63*, 114–125.

17. Xu, Z.G.; Bennett, M.T.; Tao, R.; Xu, J. China's Sloping Land Conversion Program Four Years On: Current Situation, Pending Issues. *Int. For. Rev.* **2004**, *6*, 317–326.

18. Deng, X.Z.; Huang, J.K.; Rozelle, S.; Uchida, E. Cultivated land conversion and potential agricultural productivity in China. *Land Use Policy* **2006**, *23*, 372–384.

19. Feng, Z.; Yang, Y.; Zhang, Y.; Zhang, P.; Li., Y. Grain-for-green policy and its impact on grain supply in west China. *Land Use Policy* **2005**, *22*, 301–312.

20. Xu, Z.; Xu, J.; Deng, X.; Huang, J.; Uchida, E.; Rozelle, S. Grain for green versus grain: Conflict between food security and conservation set-aside in China. *World Dev.* **2006**, *34*, 130–148.

21. Weyerhaeuser, H.; Wilkes, A.; Kahrl, F. Local impacts and responses to regional forest conservation and rehabilitation programs in China, Äôs northwest Yunnan province. *Agric. Syst.* **2005**, *85*, 234–253.

22. Le, H.D.; Smith, C.; Herbohn, J.; Harrison, S. More than just trees: Assessing reforestation success in tropical developing countries. *J. Rural Stud.* **2012**, *28*, 5–19.

23. Chokkalingam, U.; Zaizhi, Z.; Chunfeng, W.; Toma, T. *Learning Lessons from China's Forest Rehabilitation Efforts: National Level Review and Special Focus on Guangdong Province*; Center for International Forestry Research (CIFOR): Bogor, Indonesia, 2006.

24. Bennett, M.T.; Mehta, A.; Xu, J. Incomplete property rights, exposure to markets and the provision of environmental services in China. *China Econ. Rev.* **2011**, *22*, 485–498.

25. He, J.; Lang, R.; Xu, J. Local Dynamics Driving Forest Transition: Insights from Upland Villages in Southwest China. *Forests* **2014**, *5*, 214–233.

26. *Conversion of Cropland to Forests Program Plan (2001–2010)*; State Forestry Administration of the PRC: Beijing, China, 2003.

27. *State Council Notice Regarding the Improvement of the Conversion of Cropland to Forests Policy*; State Council of the PRC: Beijing, China, 2007.

28. Rozelle, S.; Brandt, L.; Guo, L.; Huang, J.K. Land Rights in China: Fact, Fiction and Issues. *China J.* **2002**, *47*, 67–97.

29. Zuo, T. Implementation of the SLCP. Implementing the Natural Forest Protection Program and the Sloping Land Conversion Program: Lessons and Policy Recommendations. In *Implementing the Natural Forest Protection Program and the Sloping Land Conversion Program: Lessons and Policy Recommendations*; Xu, J., Katsigris, E., White, T.A., Eds.; China Forestry Publishing House: Beijing, China, 2002.

30. Lamb, D. *Regreening the Bare Hills: Tropical Forest Restoration in the Asia-Pacific Region*; Springer: Heidelberg, Germany, 2010.

31. SFA. *Criterion of Ecological and Economic Trees of the Conversion of Cropland to Forest Program*; State Forestry Administration, CCFP Management Office: Beijing, China. 2001.

32. Hares, M. Perceptions of ethnic minorities on tree growing for environmental services in Thailand. In *Smallholder Tree Growing for Rural Development and Environmental Services*; Springer: Amsterdam, The Netherlands, 2008; pp. 411–425.

33. Mather, A. Recent Asian forest transitions in relation to forest-transition theory. *Int. For. Rev.* **2007**, *9*, 491–502.

34. Rudel, T.K. Tree farms: Driving forces and regional patterns in the global expansion of forest plantations. *Land Use Policy* **2009**, *26*, 545–550.

35. Meyfroidt, P.; Lambin, E.F. Forest transition in Vietnam and its environmental impacts. *Glob. Change Biol.* **2008**, *14*, 1319–1336.

36. Meyfroidt, P.; Lambin, E.F. Global forest transition: Prospects for an end to deforestation. *Ann. Rev. Environ. Res.* **2011**, *36*, 343–371.

37. Putzel, L.; Dermawan, A.; Moeliono, M.; Trung, L. Improving opportunities for smallholder timber planters in Vietnam to benefit from domestic wood processing. *Int. For. Rev.* **2012**, *14*, 227–237.

38. Xu, J. China's new forests aren't as green as they seem. *Nature* **2011**, *371*, doi:10.1038/477371a.

MDPI AG
Klybeckstrasse 64
4057 Basel, Switzerland
Tel. +41 61 683 77 34
Fax +41 61 302 89 18
http://www.mdpi.com/

Forests Editorial Office
E-mail: forests@mdpi.com
http://www.mdpi.com/journal/forests

www.ingramcontent.com/pod-product-compliance
Lightning Source LLC
Chambersburg PA
CBHW051921190326
41458CB00026B/6364